Non-Vasoactive Renal Hormones

Contributions to Nephrology

Vol. 13

S. Karger · Basel · München · Paris · London · New York · Sydney

Non-Vasoactive Renal Hormones

Volume Editors
G.M. Eisenbach and *J. Brod*, Hannover

77 figures and 10 tables, 1978

S. Karger · Basel · München · Paris · London · New York · Sydney

Contributions to Nephrology

This volume contains selected papers presented at the 5th Nephrological Symposium, Hannover, June 10–11, 1977

Vol. 10: Toxic Nephropathies
Migone, L. (ed.)
VIII + 114 p., 49 fig., 23 tab., 1978. ISBN 3–8055–2832–9
Vol. 11: Unilateral Renal Function Studies
Bianchi, C., Pisa, and Blaufox, M.D., New York, N.Y. (eds.)
X + 210 p., 77 fig., 2cpl., 23 tab., 1978. ISBN 3–8055–2858–2
Vol. 12: Vasoactive Renal Hormones
Eisenbach, G.M. and Brod, J., Hannover (eds.)
VI + 146 p., 53 fig., 14 tab., 1978. ISBN 3–8055–2839–6

Cataloging in Publication
Nephrological Symposium, 5th, Hannover, 1977. Non-vasoactive Renal Hormones / volume editors, G.M. Eisenbach and J. Brod. ––
Basel ; New York : Karger, 1978. (Contributions to nephrology ; v. 13)
Selected papers presented at the symposium, which was held June 10–11, 1977.
1. Erythropoietin – congresses 2. Vitamin D – congresses 3. Natriuresis – congresses 4. Kidney – congresses 5. Hormones congresses
I. Eisenbach, Georg M., ed. II. Brod, Jan, ed. III. Title IV. Series
Wl CO778UN v.13 WK 180 N439 1977n
ISBN 3–8055–2895–7

Printed in Switzerland by Thür AG Offsetdruck, Pratteln
ISBN 3–8055–2895–7

Contents

Contr. Nephrol., vol. 13, pp. 1–2 (Karger, Basel 1978)

Introduction

Jan Brod

Hannover Medical School, Hannover

The previous volume of the *Contributions to Nephrology* dealt with the vasoactive renal hormones. Already the adjective 'vasoactive' suggests that this is not the end to the story of the endocrine function of the kidney. And – as a matter of fact – ingenious investigations of the past 10 years have unravelled several other agents originating in the kidneys having profound effects on some of the most important functions of the body: the sodium and water balance, erythropoiesis, the calcification of the bones. All these will be discussed in our symposium.

The violent discussion whether the natriuresis following the expansion of the volume of the extracellular fluid is mediated humorally or is due purely to changes or redistribution of renal blood flow is now over a quarter of a century old and by far not solved. It is still unclear what the real role or value of the various polypeptides or other substances which have been extracted from blood or urine of diuretic animals is, and where do they originate. Prof. *Nizet* will certainly make a strong case that at least some of this activity may come from the kidney (and be another renal hormone) and Prof. *Küchel* promises to convince us that it may be renal dopamine. It is also unclear whether this or any of the previous substances is the renal humoral factor affecting the tissue or venous compliance as postulated by *Lucas and Floyer* (1973) and also by ourselves (*Brod,* 1975). The open questions obviously justify Prof. *Bourgoignie* in speaking about a 'lasting enigma'.

A defect in renal hormonal activity explains yet another enigma in renal pathophysiology; the normochromic anemia of acute and chronic renal failure.

Although part of it is still best explained by the action of the hypothetical uremic toxin on the bone marrow or on the half-life of the erythrocytes in the bloodstream, the main reason for the depressed hematopoiesis appears to be the deficiency of renal erythropoietin. Its probable storage in the granules of the juxtaglomerular apparatus and the production of the active erythropoietin II from the plasmatic α_2-globulins (proerythropoietin) by the action of the inactive renal erythropoietin I is very reminiscent of the renin-angiotensin system. Whether, beyond this similarity, some real link between these two humoral systems exists, is to my knowledge still an open problem, to which we may perhaps obtain some answers from the three protagonists in this field: *Fisher, Koch and Schulz.* A probable connection to the renal prostaglandins will be discussed by Prof. *Fisher.*

Finally, renal humoral activity has recently elucidated one more aspect of the chronic renal failure. For many years, Prof. *Stanbury* and the late Prof. *Dent* in Great Britain have been stressing that in chronic renal failure some plasmatic factor is interfering with the action of vitamin D and that this is responsible for the renal osteodystrophy. It was eventually established that the conversion of the hydrocholecalciferol to the dihydrocholecalciferol is deficient and that this conversion takes place within the kidney. Thus the kidney forms a new substance from the precursor and it is therefore fully justified if Prof. *DeLuca* speaks about the 'vitamin D endocrine system' located in the kidney.

As in the past 4 years again this time the participation of the Vth Symposium is international and I have pleasure in welcoming here in addition to speakers from the German Federal Republic active participants from eight other countries in Europe and America. I would like to thank them particularly for their readiness with which they not only accepted our invitation but with which they prepared their papers in spite of other heavy obligations. My thanks also go to all my colleagues who have assisted me with their advice and who carried on their shoulders the organization of the meeting. I would like to mention quite particularly my untiring secretary, Mrs. *Irmtraut Ahrens,* who had always all the information at her fingertips and in whose competent hands all the organizational threads were joined. The generous support of the pharmaceutical firms was already mentioned thankfully in the previous volume.

References

Brod, J.: Chronic renal parenchymal disease and hypertension. Kidney int. *8:* 235 (1975).

Lucas, J. and Floyer, M.A.: Renal control of changes in the compliance of the interstitial space: a factor in the etiology of renoprival hypertension. Clin. Sci. mol. Med. *44:* 397 (1973).

Prof. Dr. *J. Brod,* DrSc, FRCP, Medizinische Hochschule Hannover, Departement für Innere Medizin, Abteilung für Nephrologie, Karl-Wiechert-Allee 9, Postfach 61 01 80, *D–3000 Hannover 61* (FRG)

Contr. Nephrol., vol. 13, pp. 3–11 (Karger, Basel 1978)

Natriuretic Factor, a Lasting Enigma

Amnon Licht, Leon G. Fine and Jacques J. Bourgoignie

Division of Nephrology, Department of Medicine, University of Miami School of Medicine, Miami, Fla.

Introduction

In his classical review on sodium and water excretion, *Smith* (1) postulated in 1957 the existence of an antinatriuretic hormone. In 1961, *de Wardener et al.* (2) obtained evidence in dogs that a factor or factors other than glomerular filtration rate, aldosterone and vasopressin regulated the excretion of sodium by the kidney. Subsequent to *de Wardener*'s observation a large amount of experimental work in various laboratories throughout the world has been performed in an attempt to dissect out the component parts of the control system regulating sodium excretion; and a number of additional factors, which can be listed under the general categories of physical and humoral, have been shown to influence sodium handling by the nephrons (3, 4). So far, however, none of these factors has been preeminent in explaining the mechanisms whereby renal excretion of sodium is governed.

Our own efforts, initiated several years ago, have been directed at the enigma of natriuretic hormone. Using gel filtration on Sephadex G-25, we have accumulated evidence for the existence of a low molecular weight inhibitor of sodium transport in serum and urine which inhibits short-circuit current (SCC) across the isolated toad bladder and frog skin and is natriuretic in the rat (5–11). Several laboratories using the same methodology have confirmed or extended some of our observations (12–14). In brief, a low molecular weight natriuretic factor has been detected in serum and urine of man and dog under a variety of conditions associated with a high fractional excretion of sodium. By contrast, the factor has not been detectable in man or experimental animals when fractional excretion of sodium is low. Table I lists the results of a series of clinical and experimental conditions in which natriuretic factor was sought in fractions of serum and urine. Using different preparative procedures *Buckalew and Lancaster* (15), *Buckalew and Nelson* (16), *Kramer* (17), *Kramer and Krück*

Table I. Detectability of natriuretic factor

	Detectable in
1	Man with chronic uremia (5–8)
2	Dog with chronic uremia (9)
3	Normal dog during mineralocorticoid escape (10)
4	Normal dog on high sodium diet (10)
5	Normal man undergoing water immersion (11)
6	Man with primary aldosteronism (12)
7	Natriuretic man with head trauma (13)
8	Normal man on high sodium diet (14)
	Not detected in
1	Normal man on regular sodium intake (5–7)
2	Normal dog on moderate sodium intake (10)
3	Man with chronic uremia and salt retention secondary to the nephrotic syndrome (6)
4	Dog with chronic uremia subjected to 'proportional reduction of sodium intake' (9)
5	Man with primary aldosteronism after removal of the adrenal tumor (12)

(18) and *Gonick and Sandanha* (19) have also presented evidence favoring the existence of a low molecular weight natriuretic factor. Evidence that these separate laboratories are all dealing with the same substances is not currently available.

Major impediments have hampered progress in the efforts at purification and isolation of natriuretic factor. Clearly, the necessity exists for developing new and sensitive biochemical, immunologic or biologic methods capable of quantitating microquantities of natriuretic factor.

Experiments are reviewed in which natriuretic factor was assayed in the isolated perfused collecting tubule of the rabbit in an attempt to localize its site of action in the mammalian nephron. Evidence is also presented that a quantitative assay for natriuretic factor can be developed in the isolated toad bladder preparation. Finally, a new approach to the isolation of natriuretic factor is outlined.

Methods

Studies on Isolated Collecting Tubule and Toad Bladder

Quantitative 24-hour urine collections were obtained from patients with chronic renal disease. At the time of study, all patients had GFRs below 10 ml/min, and were in salt and water balance on a diet containing at least 80 mEq sodium per day.

After lyophilization, concentrated urine samples were fractionated by gel filtration at 4 °C on columns of Sephadex G-25 using 0.010 *M* ammonium acetate at pH 6.8, as previously described (6). The fraction which appeared immediately after the major peaks of sodium, urea and creatinine and which has been shown to contain the natriuretic factor was used. All fractions were rendered ammonium free before using them by evaporation at pH 9.5 under a stream of nitrogen gas (10).

The technique of isolated tubule perfusion has been described by *Burg and Orloff* (20). In the present studies cortical collecting tubules of the rabbit were isolated and perfused *in vitro.* The details of these studies have been previously described (21). The amount of urine fraction used in each perfusion experiment was derived from a 1-hour collection of urine.

In evaluating the possibility of developing a quantitative and standardized assay for natriuretic factor, the isolated toad bladder preparation was used. In these experiments, quarter bladders from large female toads, *Bufo marinus,* from the Dominican Republic were used, as previously described (7), in which SCC was equated with net transepithelial sodium transport.

In both the isolated tubule studies and the isolated toad bladder experiments, the urine fractions were dissolved in distilled water or mixed with the appropriate Ringer's solution. They were then adjusted so that Na, K, H ion concentrations and osmolality of the test solution equalled that of the control solution.

Studies on Isolation of Natriuretic Factor

New biochemical techniques with a sensitivity of the order of 10 pmol have recently been developed for the purification and isolation of vasopressin and oxytocin from individual rat pituitaries (22). We have adapted this methodology in our laboratories (23, 24).

In brief, the system uses 25 × 0.46 cm ID columns prepacked with H.70 (cation exchange) obtained from Hamilton, a solvent gradient of pyridine acetate buffers programmed with an LKB gradient mixer and an automatic discontinuous stream sampling valve connected to a fraction collector and to a detector system. Column effluent passes through the sampling valve and is directed to the fraction collector. At predetermined intervals, an aliquot of the column effluent is transferred from the sampling valve to the detector system, reacts with fluorescamine and enters the flow cell of an Aminco fluorometer where the emitted fluorescence is monitored and recorded.

In these experiments active Sephadex G-25 fractions of urine from dogs undergoing mineralocorticoid escape from the sodium retaining effect of 9α-fludrocortisone were used (10).

Results

Studies on Isolated Cortical Collecting Tubule

When urine fractions from patients with chronic uremia were added to the perfusate, no change in transtubular potential difference (PD), net fluid absorption, net or unidirectional sodium fluxes occurred. In contrast, when the same uremic fractions were added to the fluid bathing the peritubular surface of the tubule, a rapid decrease in intraluminal negativity occurred which was reversible after removal of the fraction.

As indicated in figure 1, these changes in transtubular PD were associated with a significant decrease in net sodium flux across the cortical collecting tubule. The mean control value for net sodium flux, in a lumen to bath direction, was 6.29 ± 0.89 pmol/cm/sec in the absence, and 3.21 ± 1.08 ($p < 0.001$) in the presence of the uremic fractions.

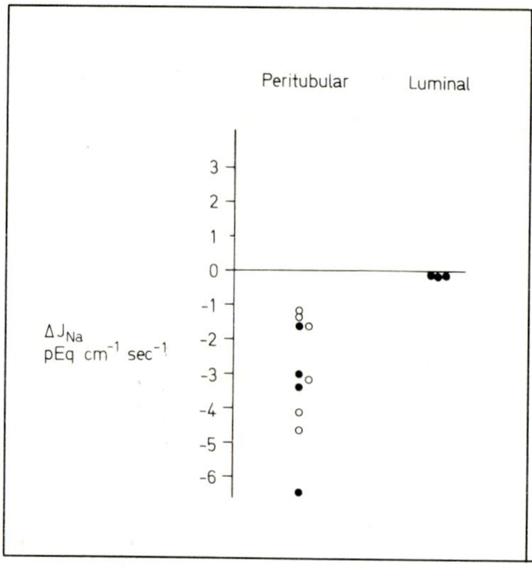

1

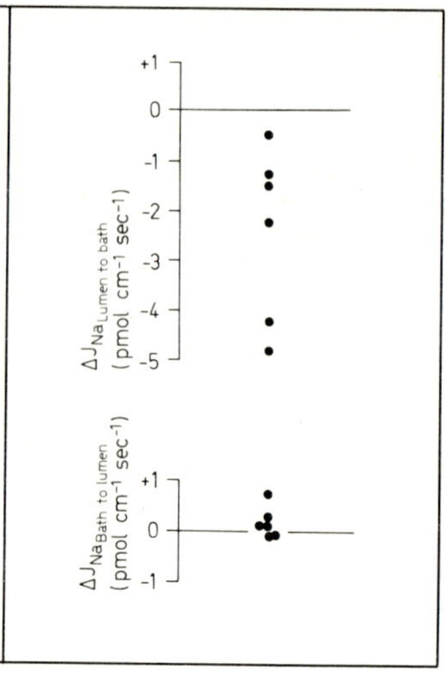

2

Fig. 1. Changes in net sodium fluxes across the isolated rabbit cortical collecting tubule after addition of natriuretic factor in the peritubular or luminal bathing solution.

Fig. 2. Changes in unidirectional sodium fluxes across the isolated rabbit cortical collecting tubule after addition of natriuretic factor in the peritubular bathing solution.

Figure 2 illustrates the effects of the uremic fractions on unidirectional sodium fluxes. The mean control value for lumen to bath flux was 8.92 ± 1.27 pmol/cm/sec. The mean value after addition of the uremic fraction was 6.31 ± 1.68 pmol/cm/sec ($p < 0.02$). The mean control value for bath to lumen flux was 1.06 ± 0.06 pmol/cm/sec and did not change after addition of the uremic fraction.

The uremic fractions had no effect on net fluid absorption. Urine fractions from normal subjects had no effect on transtubular PD or sodium fluxes of the isolated cortical collecting tubule.

Studies on Isolated Toad Bladder

Figure 3 depicts the inhibition of SCC observed with the urine fraction from a patient with chronic uremia. The same fraction was tested at different concentrations on quarter bladders from the same toad. The urine fraction, in a dose equivalent to that extracted from 0.25 h of original urine, had no effect on SCC. By contrast, after 60 min, the same fraction in amounts equivalent to 0.5, 1 and 1.5 h of original urine reduced SCC to 76, 30 and 25% of control values. The inhibition of SCC was delayed for about 20 min with the 0.5-hour dose. By contrast, the larger doses of fraction inhibited SCC within 10 min after addition of the bladder. With the three larger doses, inhibition of SCC progressed

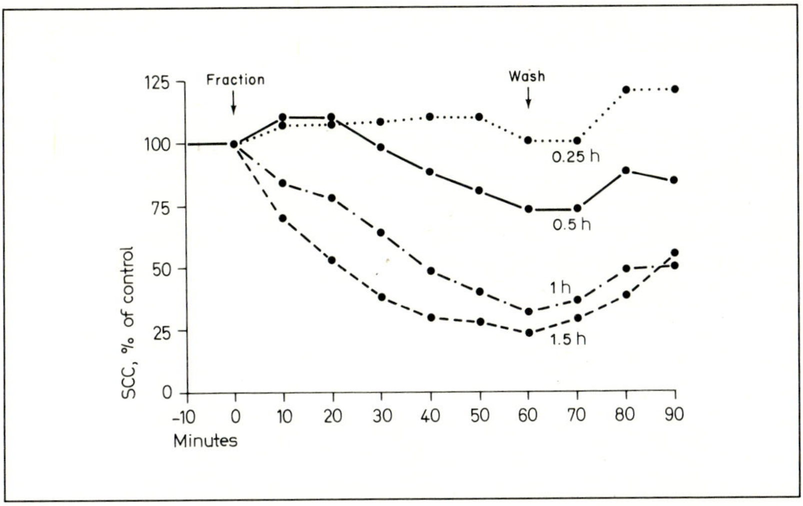

Fig. 3. Dose-response relationship between inhibition of SCC, expressed as a percent of control and concentration of urine fraction obtained from a patient with chronic uremia, expressed in hour equivalents of original urine.

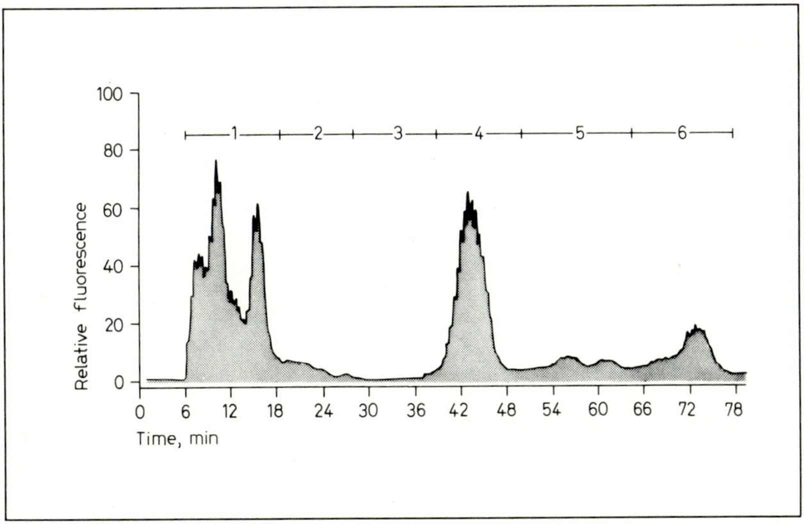

Fig. 4. HPLC was performed on a 25 × 0.46 cm ID column packed with H70 (Hamilton). A 2-hour linear gradient from $5 \times 10^{-2}\,M$ pyridine, pH 4.0 to $5 \times 10^{-1}\,M$ pyridine, pH 5.0 was used at a flow rate of 13 ml/h at room temperature with a back pressure of 300–400 psi.

throughout the 60-min period of observation. After washing the membrane, there was a return of SCC values toward control levels.

Studies on Isolation of Natriuretic Factor

The results of fractionating on a cation exchange column 0.5 ml of active Sephadex G-25 fraction, equivalent to 5 h urine of dogs undergoing mineralocorticoid escape from the sodium retaining effects of 9α-fluorocortisone are shown in figure 4. Six fractions were obtained in less than 90 min. When each of these fractions was assayed in the bladder assay, SCC inhibition consistently occurred with fraction 1 only.

Discussion

The pattern of inhibition of sodium transport by uremic fractions across the isolated cortical collecting tubule closely resembles the inhibition of SCC induced across the isolated toad bladder with the same fractions. In both preparations the inhibitory activity of the uremic fraction was evident only when the fractions were applied to the peritubular or serosal surface of the membrane; and, in neither preparation was there detectable evidence of inhibitory activity after addition of the fractions to the luminal or mucosal bathing medium. These observations are in agreement with the cumulative evidence that the toad bladder is a structure with many similarities of the distal nephron. The fact that natriuretic factor does act on the collecting tubule, the nephron segment that has recently been implicated as the final site of regulation of sodium exretion in experimental animals (25), suggests that it may serve to modulate sodium excretion by the kidney.

Any attempt at isolating and purifying a biologic substance requires that an assay preparation be available in which the activity of the biologic substance under scrutiny can be quantitated precisely. The availability of a quantitative assay is dictated not only by the need to compare the activities of different samples but to support studies on purification procedures. A standard unit of reference must also be available.

Although the action of natriuretic factor on isolated segments of rabbit cortical collecting tubule is reproducible, this preparation is too complicated and not practical for use as a routine assay. We have not attempted to determine a dose-response relationship with natriuretic factor in the isolated collecting tubule. A dose-response relationship has been demonstrated with natriuretic factor in the rat (6). However, the amount of material necessary to perform these experiments is prohibitive.

The present studies clearly demonstrate a dose-dependent inhibition of SCC across the isolated toad bladder by urine fraction from patients with chronic uremia. When quarter bladders from the same animal were tested with increasing amounts of fractions, the degree of inhibition of SCC regularly increased. We also have observed a similar dose-dependent inhibition of SCC across the toad bladder with the urine fraction from normal dogs undergoing escape from the sodium-retaining effect of mineralocorticoid hormone.

Quantitative inhibition of SCC and net sodium flux across the isolated toad bladder have previously been reported with ammonium chloride at concentrations of 0.5–10 m*M* (26). By comparing the inhibitory effects of simultaneous assays of urine fractions and of NH_4Cl in quarter bladders from the same toad, it may be possible to express quantitatively the activity of natriuretic factor.

The preliminary data on the isolation of natriuretic factor obtained with the high liquid chromatography system are very encouraging. The chromatography is fast and reproducible. The use of fluorescamine in the detector system makes the procedure very sensitive. Finally, discontinuous stream sampling of the column effluent offers the advantage of minimizing the sample consumption. Typically, 5–10% of the column eluate is used for monitoring while the remainder can be used for bioassay or other analytical purposes. The behavior of natriuretic factor in the cation-exchange system described confirms the acidic nature of the active compound.

It is our hope that the development of a quantitative assay for natriuretic factor and the availability of these new chromatographic procedures will help solve the enigma of natriuretic hormone.

Summary

A gel filtration fraction of urine from patients with chronic uremia (natriuretic factor) decreases potential difference, net sodium flux and lumen to peritubular flux of sodium across the isolated rabbit cortical collecting tubule. These effects are consistent with the possibility that natriuretic factor represents a modulator of sodium excretion in the mammalian nephron.

Natriuretic factor induces a dose-dependent inhibition of short-circuit current in the isolated toad bladder. By comparing the effects of natriuretic factor to those of a standard unit of reference, it may be possible to develop a quantitative assay for natriuretic factor. The acidic nature of natriuretic factor was confirmed by cation exchange column in a high pressure liquid chromatography system.

Acknowledgments

This work was supported by Grant AM 16281 from the National Institutes of Health, by a grant-in-aid from the American Heart Association and Florida Heart Association and by Hoechst Pharmaceutical Company. Dr. *Bourgoignie* is an Investigator of the Howard Hughes Institute for Medical Research.

References

1 *Smith, H.W.:* Salt and water receptors. An exercise in physiologic apologetics. Am. J. Med. *23:* 623–652 (1957).

2 *Wardener, H.E. de; Mills, I.H.; Clapham, W.F., and Hayter, C.J.:* Studies on the efferent mechanism of the sodium diuresis which follows the administration of intravenous saline in the dog. Clin. Sci. *21:* 249–258 (1961).

3 *Schrier, R.W. and Wardener, H.E. de:* Tubular reabsorption of sodium ion: influence of factors other than aldosterone and glomerular filtration rate. New Engl. J. Med. *285:* 1231–1244; 1292–1303 (1971).

4 *Levinsky, N.G.:* Natriuretic hormones. Adv. Metab. Disorders *7:* 37–72 (1974).

5 *Bourgoignie, J.; Hwang, K.; Espinel, C.; Klahr, S., and Bricker, N.S.:* A natriuretic factor in the serum of patients with chronic uremia. J. clin. Invest. *51:* 1514–1528 (1972).

6 *Bourgoignie, J.; Hwang, K.; Ipakchi, E., and Bricker, N.S.:* The presence of a natriuretic factor in urine of patients with chronic uremia. The absence of the factor in nephrotic uremic patients. J. clin. Invest. *53:* 1559–1567 (1974).

7 *Bourgoignie, J.; Klahr, S., and Bricker, N.S.:* Inhibition of transepithelial sodium transport in the frog skin by a low molecular weight fraction of uremic serum. J. clin. Invest. *50:* 303–312 (1971).

8 *Kaplan, M.A.; Bourgoignie, J.; Rosecan, J., and Bricker, N.S.:* The effects of the natriuretic factor from uremic urine on sodium transport, water, and electrolyte content, and pyruvate oxidation by the isolated toad bladder. J. clin. Invest. *53:* 1568–1577 (1974).

9 *Schmidt, R.W.; Bourgoignie, J., and Bricker, N.S.:* On the adaptation in sodium excretion in chronic uremia. The effects of proportional reduction of sodium intake. J. clin. Invest. *53:* 1736–1741 (1974).

10 *Favre, H.; Hwang, K.H.; Schmidt, R.W.; Bricker, N.S., and Bourgoignie, J.:* An inhibitor of sodium transport in the urine of dogs with normal renal function. J. clin. Invest. *56:* 1302–1311 (1975).

11 *Epstein, M.; Bricker, N.S., and Bourgoignie, J.:* Presence of natriuretic factor in urine of normal subjects undergoing water immersion. Kidney int. *13:* 152–158 (1978).

12 *Favre, H.:* in *Kramer and Kruck* Natriuretic hormone. Int. Symp., Bonn 1976 (in press).

13 *Klahr, S. and Rodriguez, H.J.:* Natriuretic hormone. Nephron *15:* 387–408 (1975).

14 *Clarkson, E.M.; Raw, J.M., and Wardener, H. de:* Two natriuretic substances in extracts of urine from normal man when salt-depleted and salt-loaded. Kidney int. *10:* 381–395 (1976).

15 *Buckalew, V.M., jr. and Lancaster, C.D.:* The association of a humoral sodium transport inhibitory activity with renal escape from chronic mineralocorticoid administration in the dog. Clin. Sci. *42:* 69–78 (1972).

16 *Buckalew, V.M., jr. and Nelson, D.B.:* Natriuretic and sodium transport inhibitory activity in plasma of volume-expanded dogs. Kidney int. *5:* 12–22 (1974).

17 *Kramer, H.J.:* Natriuretic hormone: its possible role in fluid and electrolyte disturbances in chronic liver disease. Post-grad. med. J. *51:* 532–540 (1975).

18 *Kramer, H.J. and Kruck, F.:* Antinatriferic activity in human plasma following acute and chronic salt load. Kidney int. (in press).

19 *Gonick, H.C. and Sandanha, L.F.:* A natriuretic principle derived from kidney tissue of volume-expanded rats. J. clin. Invest. *56:* 247–255 (1975).

20 *Burg, M.B. and Orloff, J.:* Perfusion of isolated renal tubules; in Handbook of physiology,

section 8: Renal physiology, pp. 145–159 (American Physiological Soc., Bethesda, Md. 1973).

21 *Fine, L.G.; Bourgoignie, J.; Hwang, K.H., and Bricker, N.S.:* On the influence of natriuretic factor from patients with chronic uremia on the bioelectric properties and sodium transport of the isolated mammalian collecting tubule. J. clin. Invest. *58:* 590–597 (1976).

22 *Gruber, K.A.; Stein, S.; Brink, L.; Radhakrishnan, A., and Udenfriend, S.:* Fluorometric assay of vasopressin and oxytocin. A general approach to the assay of peptides in tissues. Proc. natn. Acad. Sci. USA *73:* 1314–1318 (1976).

23 *Bohlen, P.; Stein, S.; Stone, J., and Udenfriend, S.:* Automatic monitoring of primary amines in preparative column effluent with fluorescamine. Analyt. Biochem. *67:* 438–445 (1975).

24 *Radhakrishnan, A.N.; Stein, S.; Licht, A.; Gruber, K.A., and Udenfriend, S.:* High efficiency cation-exchange chromatography of polypeptides and polyamines in the nanomole range. J. Chromat. *132:* 552–555 (1977).

25 *Stein, J.H. and Reineck, H.J.:* Effect of alterations in extracellular fluid volume on segmental sodium transport. Physiol. Rev. *55:* 127–141 (1975).

26 *Guggenheim, S.J.; Bourgoignie, J., and Klahr, S.:* Inhibition by ammonium of sodium transport across isolated toad bladder. Am: J. Physiol. *220:* 1651–1659 (1971).

Dr. *J. Bourgoignie,* Division of Nephrology, Department of Medicine, University of Miami School of Medicine, PO Box 520875, *Miami, FL 33152* (USA)

Contr. Nephrol., vol. 13, pp. 12–20 (Karger, Basel 1978)

Third Factor and Edema Formation

F. Krück and H.J. Kramer

Medizinische Universitäts-Poliklinik (Direktor: Prof. *F. Krück*), Bonn

The idea of a possible existence of an endogenous humoral natriuretic activity arose from clinical as well as from physiological considerations and studies. One of these points was the observation that cardiac patients may develop sodium retention and edema without remarkable decrease in glomerular filtration rate and/or increase in aldosterone activity. Another point was the fact that patients with Conn's syndrome do not show edema despite high aldosterone concentration as long as the function of the circulatory system is intact.

It was *Starling* (15) who pointed out in 1909 that the renal excretion of salt and water must be conditioned by the volumes of the body fluids as well as by their composition. In 1935, *Peters* (13) suggested that one factor conditioning sodium excretion would be some function of the volume of the circulating blood and specifically some change in its distribution. And still in 1957, *Smith* (14) emphasized that whereas numerous circumstances are able to induce unequivocal antinatriuresis, no physiological circumstance is known which leads to unequivocal natriuresis except the administration of sodium chloride. But salt administration would mean expansion of extracellular fluid volume.

At this time it was already shown that expansion of extracellular volume by infusions of either saline or glucose in animal and man is followed by suppression of tubular sodium reabsorption and by an increase of urinary sodium excretion (8). The factors leading to this increase of tubular sodium rejection during expansion of extracellular volume are still not entirely defined. At least two different mechanisms by which this response might be mediated have been proposed: Physical or hemodynamic factors such as changes in glomerular filtration rate, renal perfusion pressure or renal arteriolar tone, alterations of capillary absorption with consequences on the rate of epithelial transport functions or alterations of blood viscosity (8).

Although a rise in filtered load of sodium plays an important role in natriuresis resulting from sodium loading, experiments by *de Wardener et al.*

(17) demonstrated that an increase in sodium excretion can also occur after volume expansion even when the glomerular filtration rate is reduced. Therefore it was suggested that extracellular volume expansion might stimulate the production of a humoral substance reponsible for this natriuretic effect.

In the early sixties in our laboratories we were able to show that also oral hydration in normal man causes a statistically significant increase in sodium excretion without affecting tubular sodium load and without major changes in aldosterone activity or hematocrit (5, 10).

Subsequently we have shown that 25 μg of the dialyzed and lyophilized urine extract of these prehydrated persons, according to a method described by *Little et al.* (12), causes sodium and chloride diuresis together with an increase in urine volume when injected intravenously into sodium-depleted rats (6, 7, 9, 11) (fig. 1). From the decrease of negative free water clearance it was concluded that the cortical diluting segment of the nephron could be the site of action of this natriuretic activity (fig. 2) (9).

Balance studies detected that sodium-depleted rats receiving only saline retain a great portion of the given sodium and water. The animals still maintain a positive balance. On the other hand, animals treated with 25 μg of the natriuretic urine extract in combination with the same amount of saline develop a highly negative sodium and water balance (fig. 3) (6, 7, 9, 11).

To find out whether or not the urine of patients with cardiac edema also contains a natriuretic activity, urine extracts from these patients were tested in

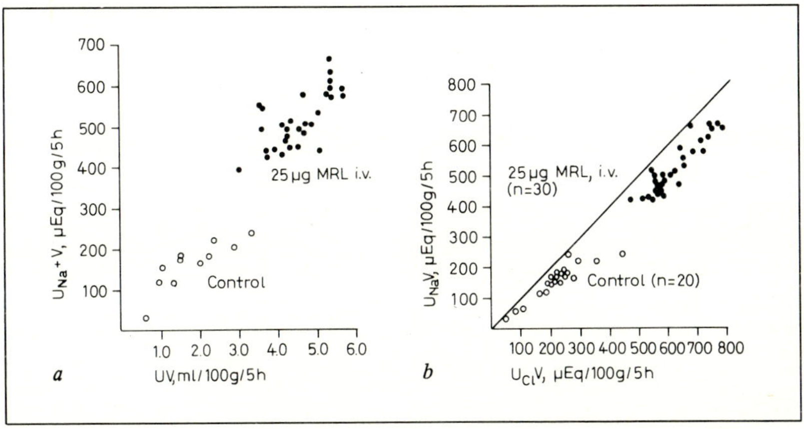

Fig. 1. *a* Relationship between sodium excretion and urine volume after injection of 25 μg of urine extract MRL (Na-depleted rats). ○ = Control experiments. *b* Relationship between sodium and chloride excretion after injection of 25 μg of urine extract MRL (Na-depleted rats). ○ = Control experiments. MRL = Membrane residue L (active natriuretic urine fraction).

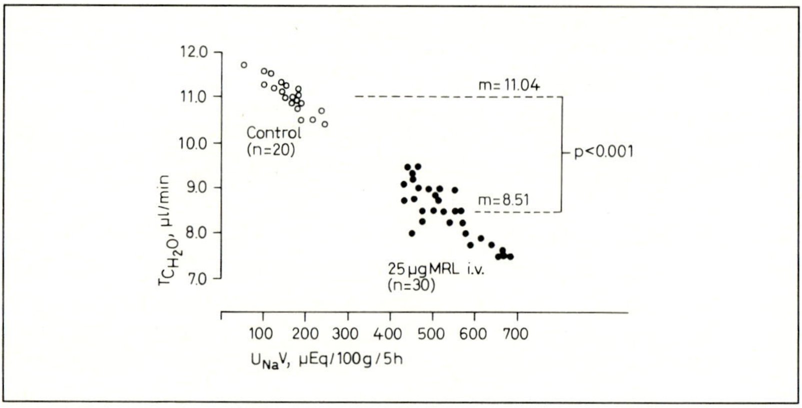

Fig. 2. Effect of 25 µg MRL on negative free water clearance and sodium excretion in sodium-depleted rats. ○ = Control experiments.

the bioassay. Instead of an exogenous water load, sodium and water retention was believed to produce approximately an equal extracellular volume expansion.

In contrast to the effect of urine extracts from normal persons after oral hydration, extracts obtained from cardiac patients had no effect on renal sodium and water excretion in sodium-depleted animals: The rats behaved as when given just saline, the sodium and water balance continued to be positive (fig. 3).

From these observations with the dialyzed and lyophilized high molecular weight natriuretic urine fraction it was concluded that patients with developing edema due to cardiac insufficiency fail to produce or to bring in action a humoral natriuretic activity.

In contrast to the effects on sodium and water excretion in rat bioassay the high molecular weight extracts had no effect on potential difference and short circuit current when applied to the inside of frog skin in the Ussing chamber (16).

Therefore, urine and plasma samples of volume-expanded healthy persons and of patients with edema of different origin were fractionated by column chromatography using Sephadex G-25, as reported by *Kramer et al.* (3). A similar method was described by *Bourgoignie et al.* (1). By this method several fractions were obtained of which fraction IV proved to have a natriuretic activity (3).

Plasma fraction IV obtained from normal persons after extracellular volume expansion showed a distinct decrease in short circuit current and potential difference when applied to the inner surface of the abdominal skin of *Rana temporaria* which was reversible by washing out with Ringer's solution (fig. 4). This points to an inhibition of sodium transport by the plasma fraction.

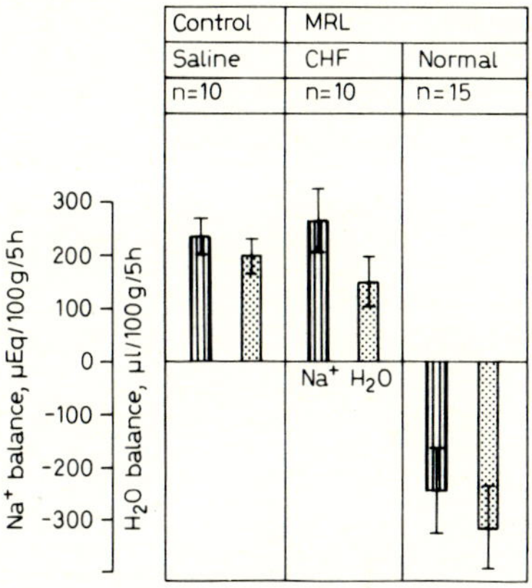

3

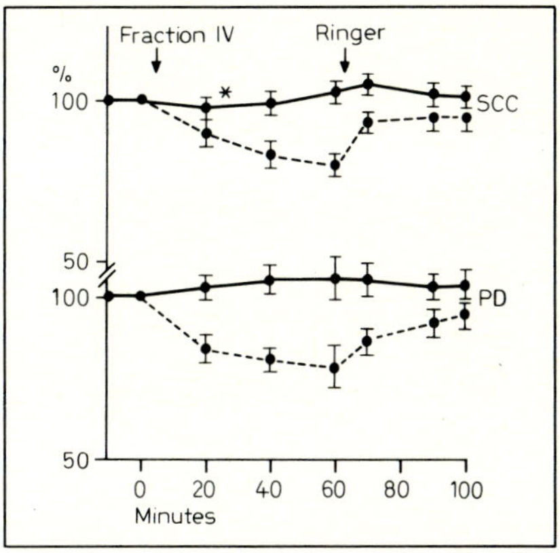

4

Fig. 3. Effect of urine extract MRL from patients with congestive heart failure (CHF) as compared with urine extract from normal persons. Left: Control experiments without MRL.

Fig. 4. Effects of plasma fraction IV of control (——) and volume expanded (- - -) normal persons on short circuit current (SCC) and potential difference (PD) of the frog skin. * = SE, 6 subjects.

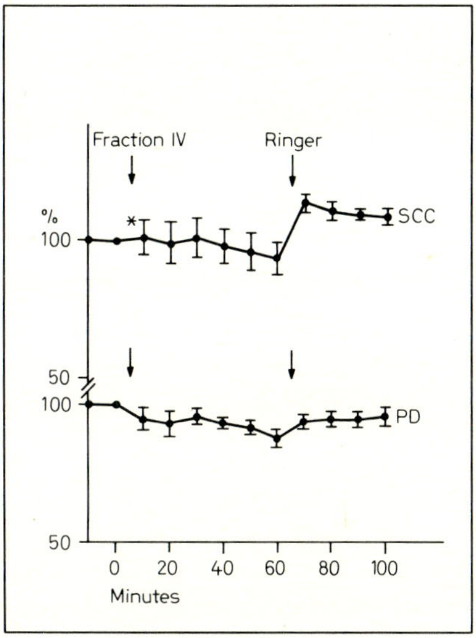

5

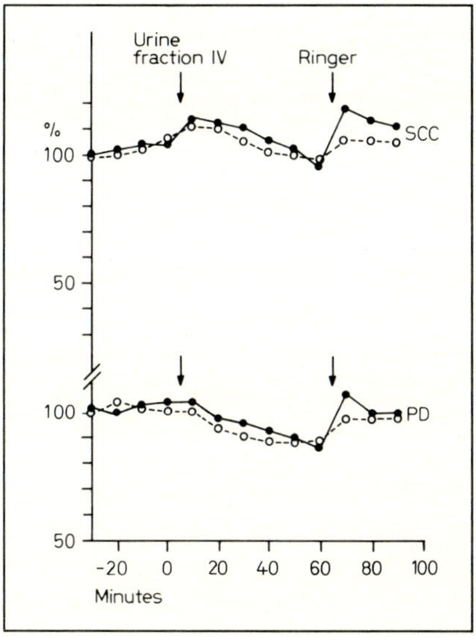

6

Fig. 5. Effect of plasma fraction IV from patients with cardiac insufficiency on SCC and PD of the frog skin. * SE, 4 subjects.

Fig. 6. Effect of urine fraction IV from cardiac patients on SCC and PD of the frog skin. —— = Patient O.H.; - - - = patient E.H.

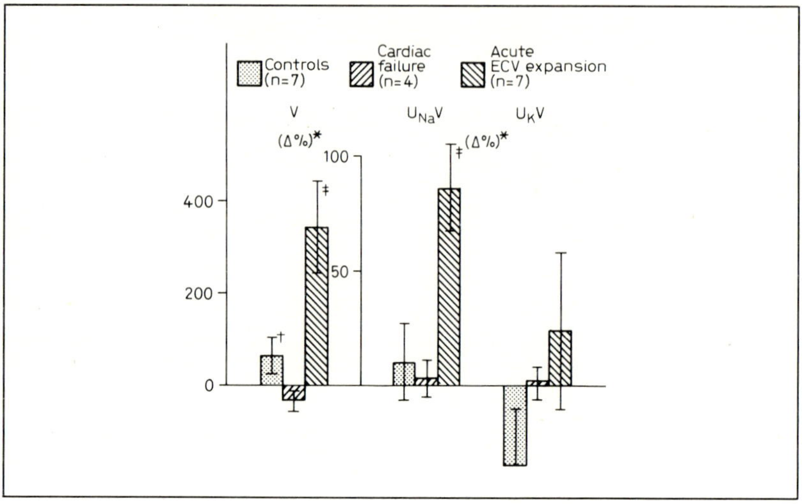

Fig. 7. Effect of plasma fraction IV on urine volume, sodium and potassium excretion. * = % change from baseline; † = SE; ‡ = $p < 0.05$. ECV = Extracellular volume.

To continue our early experiments with the high molecular weight natriuretic fraction in cardiac edema, further studies were done in patients with moderate cardiac insufficiency where the liver was clinically not affected. Plasma aldosterone concentration and plasma renin activity were measured in all but one patient and proved to be normal. All but one patient had pulmonary congestion and right-sided pleural effusion. Plasma extracts were prepared in the same manner as in normal persons. They were applied to the frog skin.

In contrast to the effect of plasma fractions obtained from normal persons with extracellular volume expansion, plasma fractions from patients with cardiac edema, whose endogenous expansion was thought to be due to an increase in interstitial fluid volume as the consequence of sodium and water retention, failed to alter short circuit current and potential difference when applied to the frog skin (fig. 5). Urine fraction IV from patients with cardiac edema also did not induce alterations of the parameters of sodium transport of the frog skin (fig. 6).

When injected into rats, plasma fractions of patients with cardiac edema also did not cause sodium diuresis, in contrast to plasma fractions from normal persons after extracellular volume expansion with Ringer's solution (fig. 7).

Negative results on short circuit current and potential difference were also obtained by using plasma fraction IV from patients with cirrhosis of the liver and nephrotic syndrome (fig. 8, 9). In rat bioassay also no effect could be observed (2).

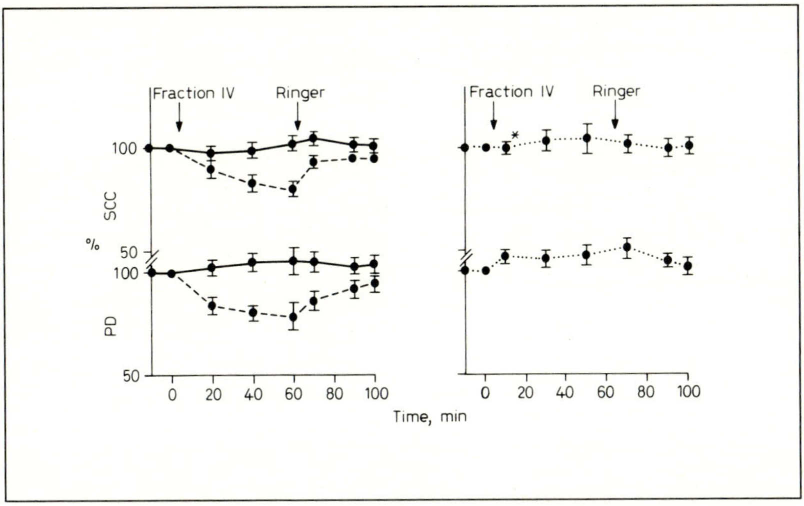

Fig. 8. Mean changes in PD and SCC of isolated frog skin induced by fraction IV of plasma from healthy subjects before and after ECV expansion and from patients with liver cirrhosis and ascites when added to the Ringer solution bathing the inner surface. —— = Controls; – – – = ECV expansion; ········ = cirrhotics. * = SE, 6 subjects.

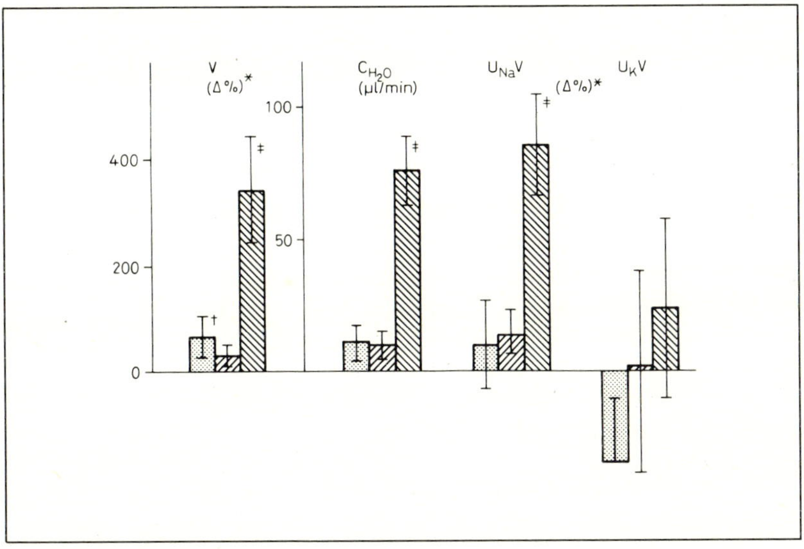

Fig. 9. Effect of plasma fraction IV obtained from normal persons with and without volume expansion and from cirrhotics (n = 6) on urine volume, free water clearance and electrolyte excretion. * = % changes from baseline; † = SE (standard error); ‡ $p < 0.05$.

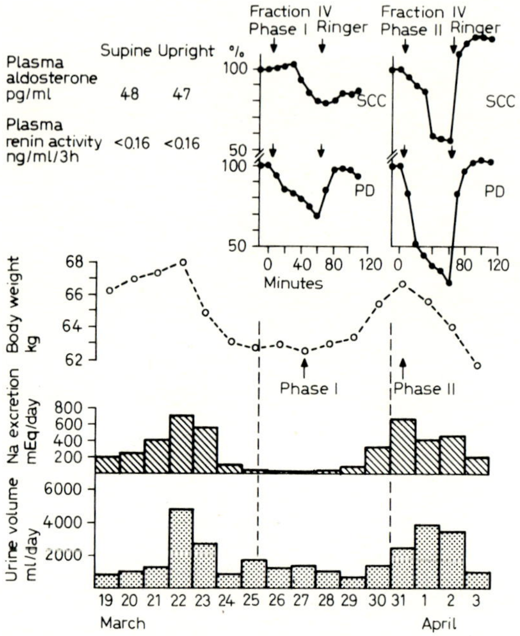

Fig. 10. Effect of plasma fraction IV obtained in different phases of idiopathic edema in a 42-year-old female patient.

In patients suffering from so-called idiopathic edema with recurrent episodes of sodium retention followed by spontaneous sodium diuresis, only minor effects of plasma fraction IV on short circuit current and potential difference occurred during retention periods whereas at the onset of natriuresis a significant inhibition of frog skin sodium transport was observed (4) (fig. 10).

Inspite of high aldosterone activity plasma fraction IV from patients with Conn's syndrome revealed a significant effect on short circuit current and potential difference when applied to frog skin which was further increased during additional salt load (fig. 11).

The results reported here show a good correlation between sodium diuresis and humoral natriuretic or antinatriferetic activity in plasma and urine which is lost in sodium-retaining states. Whether this means a causal pathophysiologic principle has to be clarified by further investigations.

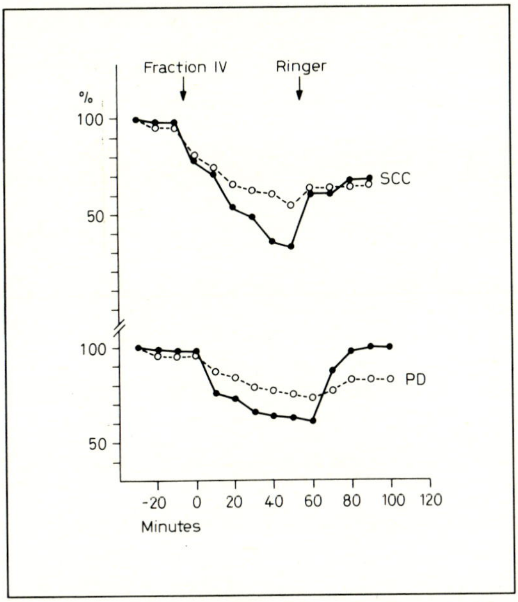

Fig. 11. Effect of plasma fraction IV from a 39-year-old female patient with Conn's syndrome before (○) and during (●) salt loading.

Summary

Urine extracts as well as plasma and urine fractions from normal persons during extracellular volume expansion and from patients with edema of different origin were tested for natriuretic (rat bioassay) and antinatriferetic (frog skin) activity. Whereas the extracts and fractions from normal persons with extracellular volume expansion proved to have a distinct natriuretic and antinatriferetic effect, no such effect was observed with urine and plasma preparations from patients with sodium retention and edema. Whether or not edema formation is mediated by a loss of natriuretic activity remains still to be clarified.

References

1 *Bourgoignie, J.; Klahr, S., and Bricker, N.S.:* Inhibition of transepithelial sodium transport in the frog skin by a low molecular weight fraction of uremic serum. J. clin. Invest. *50:* 303–311 (1971).

2 *Kramer, H.J.:* Natriuretic hormone – its possible role in fluid and electrolyte disturbances in chronic liver disease. Post-grad. med. J. *51:* 532–540 (1974).

3 *Kramer, H.J.; Gospodinow, B. und Krück, F.:* Humorale Hemmung des epithelialen Natrium-Transports nach akuter Expansion des Extracellularvolumens. Weitere Untersuchungen zur Existenz eines natriuretischen Hormons. Klin. Wschr. *52:* 801–808 (1974).

4 *Kramer, H.J. and Krück, F.:* Plasma natriuretic activity in oedematous states. Proc. Eur. Dial. Transplant. Ass. *12:* 321–329 (1976).

5 *Krecke, H.-J. und Krück, F.:* Die Natriurese bei oraler Hydratation. Verh. dt. Ges. inn. Med. *70:* 977–980 (1964).

6 *Krück, F.:* Biologischer Nachweis eines humoralen natriuretischen Prinzips im Urin gesunder Menschen. Klin. Wschr. *45:* 30–34 (1967).

7 *Krück, F.:* Influence of humoral factors on renal tubular sodium handling. Nephron *6:* 205–216 (1969).

8 *Krück, F.:* Das natriuretische Prinzip. Bedeutung und Nachweisversuche. Medsche Klin. *65:* 1155–1160 (1970).

9 *Krück, F.:* Physiological natriuretic activity in human urine; in *Cort and Lichardus* Regulation of body fluids by the kidney, pp. 100–113 (Karger, Basel 1970).

10 *Krück, F. and Krecke, H.-J.:* The renal sodium excretion during oral hydration in man. Nephron *2:* 321–333 (1965).

11 *Krück, F. und Torbica, M.:* Untersuchungen über die natriuretische Wirksamkeit verschiedener Urinfraktionen. Verh. dt. Ges. inn. Med. *72:* 391–394 (1966).

12 *Little, J.M.; Angell, E.A., and Brooks, W.:* Diuretic activity of three primary fractions of total nondialyzable solids from human urine. Proc. Soc. exp. Biol. Med. *111:* 316–320 (1962).

13 *Peters, J.P.:* Body water. The exchange of fluids in man (Thomas, Springfield 1935).

14 *Smith, H.W.:* Salt and water volume receptors. Am. J. Med. *23:* 625–652 (1957).

15 *Starling, E.H.:* The fluids of the body; in Herter lectures 1909, p. 106 (Keener, Chicago 1909).

16 *Stumpe, K.O.; Pöhler, E., and Krück, F.:* Diuretic activity in human urine. Lack of action on proximal tubule and frog skin sodium transport. Res. exp. Med. *162:* 323–332 (1974).

17 *Wardener, H.E. de; Mills, I.H.; Clapham, W.F., and Hayer, C.J.:* Studies on the efferent mechanism of the sodium diuresis which follows the administration of intravenous saline in the dog. Clin. Sci. *21:* 249–258 (1961).

Dr. *F. Krück*, Medizinische Universitäts-Poliklinik Bonn, *Bonn* (FRG)

Contr. Nephrol., vol. 13, pp. 21–26 (Karger, Basel 1978)

Humoral Control by the Kidney of Intestinal Transport of Sodium

A. Nizet, M. Robin, G. Merchie and J.P. Godon
Institute of Medicine, Hôpital de Bavière, Liège

Our previous experimental results seem to indicate that a specific natriuretic hormone of extrarenal origin plays little, if any, role in the fast adjustments of fractional excretion of sodium which follow an acute intravenous saline loading. A totally isolated kidney excretes quantitatively an excess of saline added to the perfusion blood (10); moreover, the natriuretic response is equal in a totally isolated kidney and in a kidney transplanted to the neck vessels of a perfusor dog provided that the degree of blood dilution is the same (9). No relationship can be found with the changes occurring in either intravascular and interstitial volume spaces (12). The changes in overall fractional reabsorption of sodium are proportional to the direct cumulative effect on the kidney of dilution indexes such as plasma oncotic pressure, hematocrit and ionic concentration (8). However, the renal response to blood dilution is considerably reduced in transplanted (12) as well as totally isolated (9) kidneys taken from sodium-depleted dogs, thus indicating an adjustment of the sensitivity of the kidney to the blood dilution indexes.

In the following discussion, the dogs submitted to a sodium-free diet will be called 'Na–' dogs, and the animals submitted to a sodium-enriched diet will be called 'Na+' dogs.

A 'Na+' kidney transplanted on the neck vessels of a 'Na–' dog exhibits a maximal natriuretic response to the intravenous infusion of saline; moreover, the response of the 'Na–' kidney *in situ* is potentiated; thus the 'Na+' kidney bears the natriuretic message and releases it into the blood (12). This interpretation is confirmed by the detection of a natriuretic material in the blood perfusing a totally isolated 'Na–' kidney after addition of a saline load *in vitro* (7). Similar conclusions concerning the renal origin of a plasma natriuretic material were reached by one of us by using the pathological model of experimental glomerulonephritis in the rat (3, 4). It appeared therefore that a primary determinant of a saline natriuresis is the cumulative effect on the kidney itself of blood dilution indexes. However, the response of the kidney depends on the previous dietary sodium balance and on the presence in this organ of one or several factors, some of which, at least, are released into the blood and appear as a secondary response to the direct renal effect on blood dilution or chronic sodium loading.

Some similarities exist between the proximal tubule and the jejunum as concerns water and sodium transport. In both structures, an acute saline loading decreases net sodium and water absorption (5, 6, 13) and a distension of intercellular spaces is observed (2). An increased flow from intercellular channels into the intestinal lumen might be primarily due to the decrease of plasma oncotic pressure (6). This mechanism is quite similar to the one described in the proximal tubule (1).

Another interpretation was suggested by *Richet and Hornych* (13): the inhibition of sodium and water absorption in both the proximal tubule and the jejunum might be due to the presence in the blood of a natriuretic factor acting on both organs.

It seemed logical to investigate further the characteristics of the jejunal response to the acute saline load and to evaluate the role of expansion of interstitial and intravascular volume spaces, blood dilution indexes, 'third factor' and previous dietary intake of sodium. The experimental results demonstrate a striking similarity between renal and jejunal responses (11): (a) the impairment of net absorption of water and sodium depends on the previous intake of sodium; it is not observed in a sodium-depleted dog despite equal changes in volume spaces and blood dilution; (b) the intestinal response reappears in a 'Na–' animal after transplantation to its neck vessels of a 'Na+' kidney.

Therefore, the sensitivity of the jejunum as well as of the kidney to the dilution of the blood depends on an intrarenal mechanism involving the release by the kidney into the blood of a material which modulates the response of both organs to the direct effect of the 'physical factors'. The changes in unidirectional fluxes of sodium were investigated; the difference of response between 'Na+' and 'Na–' preparations depended mainly on the mucosa-to-serosa sodium flux which was impaired by saline loading only in 'Na+' animals or in 'Na–' animals with transplanted 'Na+' kidneys, while the serosa-to-mucosa unidirectional flux was reduced in all preparations (11). All experiments were performed in dogs loaded with desoxycorticosterone acetate; a possible interference of this hormone was therefore eliminated. The potentiation of the response of a 'Na–' dog by a 'Na+' kidney seems barely compatible with a role of angiotensin.

The following experiments seem to exclude a possible role of changes in body fluid volumes as well as of a natriuretic hormone of extrarenal origin.

Totally isolated jejunum segments were perfused with heparinized blood by the pump-oxygenator system used for kidney perfusions. A cannula was inserted into the mesenteric artery and the effluent of venous blood was collected freely in a funnel and sent back to the machine. Arterial pressure was kept at a constant level. Jejunal exchanges of sodium and water were measured by the technique used in the jejunum *in situ* (11). A Ringer solution was injected into the jejunum segment at a constant flow and the differences between input and output of water were measured with the help of a non-absorbed material (polyethylene glycol). Net fluxes of sodium and water were calculated by difference; moreover, a tracer dose of ^{22}Na was added in order to measure the unidirectional fluxes of this ion (11). In a first period, jejunal transport was measured during perfusion with undiluted blood. An isotonic saline load was then added to the blood in order to obtain a degree of blood dilution comparable to the one observed in the experiments *in vivo.* After a period of equilibration, the exchanges were measured during 2–4 periods of 20 min each. Three series of experiments were performed: (a) 'Na+' jejunum perfused with 'Na+' blood; (b) 'Na–' jejunum perfused with 'Na–' blood; (c) 'Na–' jejunum perfused with 'Na–' blood, a 'Na+' kidney being connected to the oxygenator and perfused by the same blood pool with an independent pump.

The results can be summarized as follows:

(a) Contrary to the jejunum *in situ,* there is a net secretion of sodium and water instead of a net absorption. The unidirectional serosa-to-mucosa sodium flux is higher and the mucosa-to-serosa sodium flux is lower.

(b) After addition of saline, the serosa-to-mucosa unidirectional flux of sodium is increased in the three series. The unidirectional mucosa-to-serosa flux does not change in the 'Na–' jejunum. However, the introduction of a 'Na+' kidney into the perfusion system restores the response of the 'Na–' jejunum, the decrease of mucosa-to-serosa flux being equal to the one observed in the 'Na+' jejunum (table I). The impairment of mucosa-to-serosa flux is somewhat more marked in the isolated preparation. This difference might be explained by the

Table I. Changes induced by saline loading in sodium and water transport in the isolated dog jejunum

	'Na+' jejunum	'Na–' jejunum	'Na–' jejunum and 'Na+' kidney
Water secretion, μl/min/g	3.32 ± 0.84	1.96 ± 0.49	3.56 ± 0.39
		← 2 p = 0.02 →	
Net Na secretion, μEq/min/g	+ 0.573 ± 0.122	+ 0.242 ± 0.060	+ 0.515 ± 0.053
		← 2 p = 0.005 →	
Serosa-to-mucosa Na flux, μEq/min/g	+ 0.339 ± 0.118	+ 0.263 ± 0.084	+ 0.333 ± 0.062
Mucosa-to-serosa flux, μEq/min/g	– 0.174 ± 0.016	+ 0.021 ± 0.038	– 0.182 ± 0.049
	← 2 p <0.001 →	← 2 p <0.001 →	

Table II. Comparison of changes in mucosa-to-serosa Na flux induced by saline load (μEq/min/g)

	Jejunum *in situ*	Isolated jejunum
Between 'Na+' and 'Na−' preparations	0.109 ± 0.035	0.195 ± 0.041
Between 'Na−' preparations with and without addition of 'Na+' kidneys	0.116 ± 0.048	0.203 ± 0.042
Between 'Na+' preparations and 'Na−' preparations with 'Na+' kidneys	0.007 ± 0.031	0.008 ± 0.052

Table III. Relationship between mucosa-to-serosa flux of sodium and glucose absorption by dog jejunum (*in situ* or isolated)

Without saline load		
In situ	y = 21.33 (± 4.54) x + 23.20	2 p <0.001
Isolated	y = 24.78 (± 4.83) x + 3.39	2 p <0.001
Changes induced by saline load		
In situ + isolated	y = 27.11 (± 3.65) x + 1.75	2 p <0.001
Overall values (with and without saline load)		
In situ	y = 17.89 (± 4.81) x + 25.98	2 p <0.001
Isolated	y = 18.11 (± 3.91) x + 7.55	2 p <0.001

x = Mucosa-to-serosa Na flux (mEq/min/g); y = glucose absorption (vM/min/g).

fact that the dilution of the blood *in vivo* is progressive while it reaches immediately a steady state in the experiments *in vitro* (table II).

Some additional information is provided by these experiments on the isolated jejunum. The direct effect of blood dilution is demonstrated; an extra-renal natriuretic stimulus is excluded as well as a possible role of volume expansion and of antidiuretic hormone.

In the experiments performed *in situ,* the concentration of sodium was lower in the perfusion fluid than in the blood, suggesting an involvement of active sodium transport. We have measured the jejunal absorption of glucose in all experimental conditions. 'Na+' dogs, 'Na−' dogs, 'Na−' dogs with transplanted 'Na+' kidney, jejunum *in situ* or isolated, without or with saline load. No correlation at all was found between either net sodium transport and serosa-to-mucosa sodium flux. On the contrary, a highly significant positive correlation was found between glucose absorption and unidirectional mucosa-to-serosa flux of sodium, the slope of the regression line being equal in all experimental conditions (table III). These results suggest the existence of a link between both jejunal transports and provide some additional reason to believe that active sodium transport is modulated.

A natriuretic material can be extracted from the diluted blood perfusing a 'Na–' kidney. This material exhibits a molecular weight comprized between 10,000 and 50,000 and seems to be of protidic nature (7); it has also been detected in the present experiments involving the simultaneous perfusion of a 'Na–' jejunum and of a 'Na+' kidney. This material is likely to represent at least one of the humoral factors which control the jejunal as well as the renal response.

These experimental results lead to the conclusion that the modulation of sodium transport in both the proximal tubule and the jejunum following acute changes in blood concentration depends on intrarenal factors which appear in the circulating blood. It seems that we have to deal with an additional endocrine function of the kidney which tends to keep constant the concentration of the blood by purely autonomic mechanisms. The fact that the excretory response of the kidney as well as sodium transport in the jejunum (and perhaps elsewhere) depends on the adaptation of the kidney according to the previous sodium balance, represents an additional homeostatic mechanism under autonomic control of this organ.

Summary

The impairment of water and sodium absorption by the jejunum following an acute saline load depends on the dietary balance of sodium, the mucosa-to-serosa unidirectional flux of this ion being modulated by a direct effect of the changes in blood concentration and by a humoral material of renal origin which itself is released as a consequence of either chronic or acute sodium loading. A possible interference of mineralocorticosteroids, angiotensin or antidiuretic hormone seems to be excluded. A positive correlation exists between mucosa-to-serosa sodium flux and glucose absorption. The kidney controls the exchanges of sodium and water in the jejunum as well as in the kidney itself by the way of an endocrine function.

References

1 *Boulpaep, E.L.:* Permeability changes of the proximal tubule of Necturus during saline loading. Am. J. Physiol. *222:* 517–531 (1972).

2 *Di Bona, D.R.; Chen, L.C., and Sharp, G.W.G.:* A study of intercellular spaces in the rabbit jejunum during acute volume expansion and after treatment with cholera toxin. J. clin. Invest. *53:* 1300–1307 (1974).

3 *Godon, J.P.:* The site of sodium retention in glomerulonephritis. Role of a natriuretic factor of renal origin. Dialysis, transplantation and nephrology. Proc. Eur. Dial. Transplant. Ass. *13:* 520–527 (1976).

4 *Godon, J.P. and Mahieu, P.: In vitro* production of a natriuretic material by tubular cells isolated from normal and glomerulonephritic rat kidneys. VIth Int. Congr. Nephrology, Florence 1975, Abstr. 137.

5 *Higgins, J.T., jr. and Blair, N.P.:* Intestinal transport of water and electrolytes during extracellular volume expansion in dogs. J. clin. Invest. *50:* 2569–2579 (1971).

6 *Humphreys, M.H. and Earley, L.E.:* The mechanism of decreased intestinal sodium and water absorption after acute volume expansion in the rat. J. clin. Invest. *50:* 2355–2367 (1971).

7 *Godon, J.P. and Nizet, A.:* Release by isolated dog kidney of a natriuretic material following saline loading. Archs int. Physiol. Biochim. *82:* 309–311 (1974).

8 *Nizet, A.:* Excretion of sodium and water by kidneys *in situ* and by transplanted kidneys following isotonic, hypotonic, isooncotic and hyperoncotic intravenous infusions in sodium-deprived and sodium-loaded dogs. Archs int. Physiol. Biochim. *84:* 997–1015 (1976).

9 *Nizet, A.:* Comparative evaluation of fractional excretion of sodium following saline infusion in transplanted kidneys and in isolated perfused kidneys in conditions of previous high or low dietary sodium intake. Pflügers Arch. *361:* 121–126 (1976).

10 *Nizet, A.; Godon, J.P., and Mahieu, P.:* Quantitative excretion of water and sodium load by isolated dog kidney; autonomous renal response to blood dilution factors. Pflügers Arch. ges. Physiol. *304:* 30–45 (1968).

11 *Nizet, A.; Robin, M., and Merchie, G.:* Renal control of intestinal sodium and water transport in the dog. Pflügers Arch. *364:* 59–64 (1976).

12 *Nizet, A.; Tost, H., and Foidart-Willems, J.:* The control of sodium excretion following saline infusion in dogs. Pflügers Arch. *350:* 287–298 (1974).

13 *Richet, G. and Hornych, A.:* The effect of an expansion of extracellular fluids on net Na flux in the jejunum of rats. Nephron *6:* 365–378 (1969).

Prof. *A. Nizet,* Institute of Medicine, Hôpital de Bavière, Bvd de la Constitution, 66, *B–4020 Liège* (Belgium)

Contr. Nephrol., vol. 13, pp. 27–36 (Karger, Basel 1978)

Dopamine-Sodium Relationship: Is Dopamine a Part of the Endogenous Natriuretic System?[1]

Otto Kuchel, Nguyen Thanh Buu and Thomas Unger

Clinical Research Institute, Hôtel-Dieu Hospital, University of Montreal, Montreal, Que.

Substances with renal vasodilating activity are plausible candidates for a role as natriuretic hormones. In some cases they are released in response to volume expansion and related to the sodium balance (1) and their action could be explained by a combined hemodynamic and natriferic action in the kidney. Some of them, produced in the kidney, such as kallikrein (2) and renal prostaglandins (3) were considered for such a role.

The role of dopamine (DA) has not been explored because of uncertainty of the origin of urinary DA and difficulties in determining plasma DA. In pharmacological doses, DA is a potent natriuretic and renal vasodilating agent (4). Some previously observed correlations between urinary sodium and DA under different conditions such as change in posture (5), sodium loading (6, 7) or DA synthesis inhibition (8) also suggested a physiological natriuretic role of endogenous DA. The mechanism of the natriuretic effect of DA are multiple: it increases the renal plasma flow and GFR, induces a redistribution of the intrarenal blood flow (9) and inhibits the tubular transport of sodium (10). Previous experiments have shown that DA infusion increases urinary cyclic AMP excretion in man and this effect parallels the increased tubular sodium rejection (11). Since β-adrenergic stimulation and cyclic AMP have been shown to induce a decrease in the proximal tubular reabsorption of sodium (12), it is probable that the natriuretic effect of DA is a combination of its action via β-adrenergic receptors and via more specific dopaminergic receptors responsible for renal vasodilatation (4). It has been shown, for instance, that infusion of saline in man resulted in an increase in urinary free DA while urinary norepinephrine (NE) and

[1] This work was supported by a grant from the Medical Research Council of Canada to the multidisciplinary hypertension research group (*D.R.S.J. Genest, R. Boucher, W. Nowaczawski* and *O. Kuchel*).

plasma dopamine-β-hydroxylase decreased (6). It remains open whether the increase of sodium excretion is only due to a decreased sympathetic tone or also to an increased DA action.

The purpose of the present study was to examine: (1) Whether saline-induced increase of DA excretion is specifically linked to sodium or is it only a nonspecific effect of the volume expansion (e.g. by albumin)? (2) What is the source of the urinary free DA and where does its saline-induced increase originate? (3) Is there any relationship between DA excretion and sodium retention in idiopathic edema?

With respect to the first question we have found in experiments described elsewhere (13) that only saline expanded dogs had an increase in the excretion of free DA; this increase originated from the kidney as reflected by an increase in the renal clearance of free DA. Dogs expanded with hyperoncotic albumin on the other hand behaved similarly as sham-operated dogs showing no increase in their free DA excretion; albumin and saline expanded dogs had however a comparable decrease in urinary and plasma NE + epinephrine (E). The increase in the DA excretion correlated positively with the increase in the TRF_{Na} (tubular rejection fraction of sodium) in all dogs (saline expansion, albumin expansion and control), again indicating a relationship between urinary DA and sodium excretion. The increase in urinary free DA following sodium-induced volume expansion was delayed when compared to the decrease in urinary NE + E. This time dissociation of both responses is similar to that observed previously in human experiments (6) and indicates that the DA increase is not a 'spillover' effect of the inhibition of the sympathetic nervous system but an independent response. It can be concluded from these experiments that the urinary free DA increase following saline is an effect related to salt rather than to a nonspecific volume expansion.

As to the second question regarding the source of the free DA excretion, there is only very indirect evidence that dopaminergic nervous terminals exist in the kidney (14). We have found, using a new method for the hydrolysis of conjugated catecholamines (15), that plasma DA is almost entirely conjugated, and free plasma DA is undetectable in 95% of cases in man, dog and rat. There are however differences in the urinary free DA and its degree of conjugation in the three species (16). The measurements of plasma, renal tissue and urinary free and conjugated DA in rats have shown (fig. 1) that DA is almost completely conjugated not only in the plasma but also in renal tissue, while urinary DA is approximately 83% conjugated. These patterns of conjugation differ from those of NE and E. In dogs, concomitant measurements of free and conjugated DA in renal artery, renal vein and urine have shown that DA is over 90% conjugated in the artery while the blood outflowing from the kidney in the renal vein showed 15% free DA. In the urine DA was approximately 49% free and 51% conjugated (17) (fig. 2).

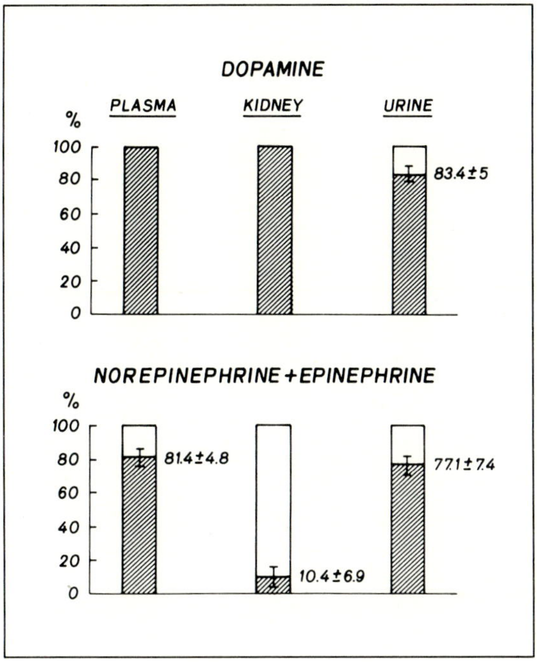

Fig. 1. Free and conjugated catecholamines in plasma, kidney tissue and urine in the rat. The conjugated catecholamines are expressed in percent of total catecholamines (mean ± SEM) by the stripped bars, the open parts of the bars are the free catecholamines. DA in plasma and in the kidney tissue occurs exclusively conjugated. Evidently different are free and conjugated NE + E. Further details in *Buu et al.* (16).

These data indicate that DA entering the kidney is almost entirely conjugated, which implies that the conjugation must occur in other tissues (brain, gut, liver) and kidney tissues contain almost exclusively conjugated DA. It is not yet clear by what mechanism the observed increase of free DA in the renal vein and urine occurs. So far, no enzymatic change responsible for increased free DA synthesis or decreased DA metabolism following sodium loading could be detected (7).

With respect to the third question we have studied the relationship between DA excretion and renal sodium handling in a puzzling salt-retaining condition, 'idiopathic' edema. This very common type of edema without underlying hepatic, cardiac or other disease was never entirely explained by known salt-retaining mechanisms particularly since extremely rapid variations in the salt balance (sometimes within hours) often occurred after emotional stress (18). Higher sympathetic activity has been postulated in these patients (19). We could

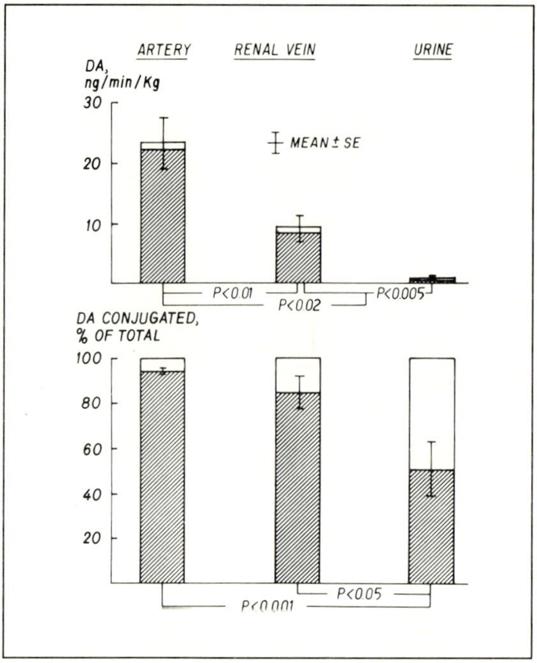

Fig. 2. Arterial, renal vein and urinary free (open bars) and conjugated (stripped bars) dopamine (mean ± SE) in 10 dogs with peroperatory sampling of blood and urine. In the upper panel are the absolute values all expressed in ng/min/kg of body weight; the arterial and renal vein values are related to the renal blood flow concomitantly measured with an electromagnetic flowmeter. The urinary values are expressed per urinary flow and per kilogram of body weight. In the lower panel are the same values but expressed as percent of the conjugated DA from the total DA (free + conjugated). Further details in *Unger et al.* (17).

not confirm the finding that these patients have higher NE excretion but found that they excrete significantly less free DA and less sodium than control subjects (20). This seems to suggest a deficiency of the dopaminergic system or a deficiency of the renal release of free DA to facilitate sodium excretion. The excessive sodium retention in 'idiopathic' edema may thus be due to a weakening of natriuretic forces normally balancing the antinatriuretic action of the sympathetic discharge towards the kidney (21). Since DA infusion inhibits renin release in salt-retaining conditions (22) and dopaminergic stimulation by bromocriptine inhibits aldosterone secretion independently of renin (23), a lack of DA action may in turn disinhibit the renin-angiotensin-aldosterone system; in fact it has been shown that idiopathic edema patients have higher renin and aldosterone responses (20). *L*-Dopa or DA induce a potent natriuresis (4, 24).

An infusion of DA normally increases nephrogenous cyclic AMP and urinary sodium excretion (11). In idiopathic edema, in contrast, a decrease of nephrogenous cyclic AMP together with the decreased urinary sodium excretion (25) probably reflects a deficiency in endogenous DA.

One of the prominent features of the secondary hyperaldosteronism accompanied by edema is the 'failure' to escape from the sodium-retaining action of aldosterone (26). On the other hand patients with secondary hyperaldosteronism and hypertension show this escape phenomenon. We compared the DA and aldosterone profiles in these two groups of patients with hyperaldosteronism and found that the DA excretion is higher in hyperaldosteronism with hypertension (be it primary aldosteronism or secondary hyperaldosteronism in renovascular hypertension) than in the normotensive hyperaldosteronism of idiopathic edema. Moreover we found that in idiopathic edema there is a positive correlation between urinary sodium excretion and the ratio of urinary DA to plasma aldosterone (27). Thus the sodium retention in idiopathic edema may be at least partly related to the balance between DA and aldosterone. In hypertensive hyperaldosteronism however, sodium retention does not occur in spite of a lower DA/aldosterone ratio indicating that probably other factors, particularly the hypertension itself, are responsible for the presence of the escape.

How does DA fit into the overall picture of the balance in which the natriuretic forces are counteracting the action of several sodium retaining factors? DA is certainly not the single elusive natriuretic hormone and it is questionable whether such a single natriuretic hormone exists at all. It is rather a part of multiple natriuretic forces acting on intrarenal physical factors and the glomerulotubular balance as well as the tubular sodium transport. Their relative weakening in idiopathic edema may explain the failure to escape from the mineralocorticoid action. Patients with primary aldosteronism have particularly high free DA excretion (28) and no sodium retention in spite of the hyperaldosteronism. This may contribute to their 'escape' at variance with idiopathic edema. Conversely the 'escape' from mineralocorticoid excess in condition of extracellular volume expansion in chronic renal insufficiency may well be related to a recent observation that uremic patients have high free plasma DA concentrations (29) contrary to control subjects who have usually undetectable plasma-free DA. DA may thus be a natriuretic hormone of prerenal (cerebral?) origin. The brain is the main site of dopaminergic nervous terminals and DA conjugation (30) and this may well be a form of circulating DA protected from being metabolized before reaching the kidney site of action.

In spite of physiological and clinical evidence favoring the role of DA as an endogenous natriuretic hormone much remains to be learned. The search for a natriuretic hormone did not traditionally consider catecholamines and renal vasodilating substances (1). The fact that urinary DA increases only during

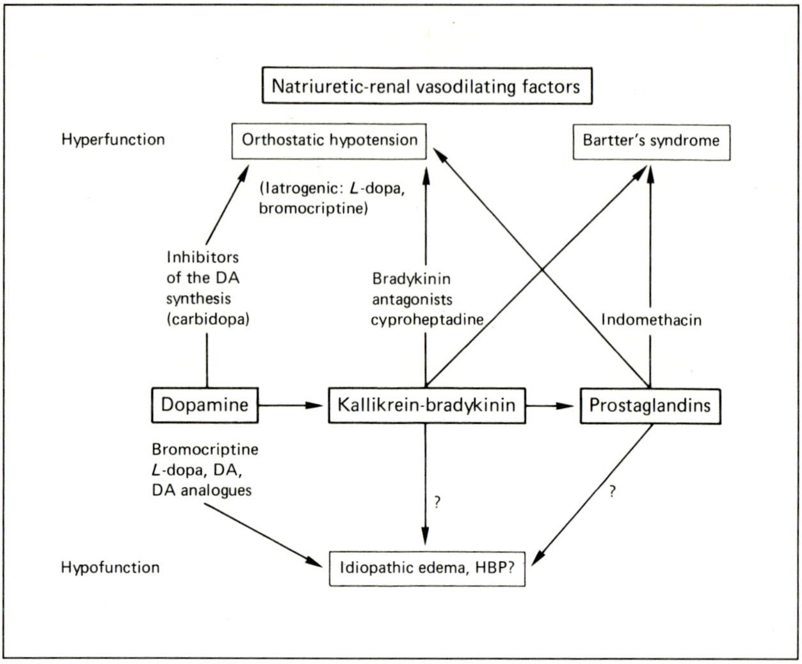

Fig. 3. Some relationships between the natriuretic-renal vasodilating factors and clinical syndromes. Some hyperfunctional syndromes are improved by inhibitors of the synthesis or action of these factors, during their deficiency (lower part) dopaminomimetic treatment appears to be of some clinical value.

salt-induced volume expansion and is not responsive to albumin expansion, links DA more specifically to salt than to volume regulation. The finding that DA excretion increases in response to furosemide administration in parallel with sodium excretion (20) speaks against the homeostatic purpose of a natriuretic hormone, which should increase in response rather to volume expansion, unless this finding reflects the dominance of sodium over volume dependency of the excretion of free DA. It is however interesting to note that the same occurs in response to furosemide, to renal prostaglandins (31) and kallikrein (32). All could be due to a 'wash out' effect of these intrarenally generated substances by furosemide-induced diuresis. Whatever the explanation, the fact remains that idiopathic edema patients with DA hypoexcretion were also found to have lower sodium excretion and lower response of DA to furosemide (20), again linking the defect of sodium excretion with a decrease in the urinary free DA excretion.

In the further search for the role of DA in the endogenous natriuretic system measurements of free DA will have to be complemented by determinations of DA-sulfate since DA is a highly conjugated catecholamine (15) and the degree of conjugation probably affects its biological activity.

In conclusion, it is suggested that DA is not the sole controlling mechanism of natriuresis but probably a part of a jointly acting natriuretic and renal vasodilating system involving renal prostaglandins, DA and the kallikrein-bradykinin system (probably also acetylcholine). Under different physiological circumstances each component of this system may have a particular role in the control of sodium excretion. DA seems to be particularly closely linked to sodium while others may have an additional effect on vascular reactivity. In clinical terms (fig. 3) dopaminergic insufficiency may contribute to the sodium retention of patients with idiopathic edema and be partially involved jointly with renal prostaglandins and kallikrein deficiency in the mechanism of some forms of hypertension (28). Dopaminergic stimulation by bromocriptine has been found to improve premenstrual idiopathic edema (33) and some forms of hypertension (34) and *L*-dopa to have a moderate hypotensive effect (35). Hyperfunction of this system involves probably renal prostaglandins as in the Bartter's syndrome (36) and orthostatic hypotension (37). The latter is in some cases also due to DA excess, be it endogenous (38) or iatrogenic (after *L*-dopa or bromocriptine) or due to hyperbradykininism (39). The successful use of inhibitors of synthesis or of antagonists of these substances such as indomethacin in Bartter's syndrome (36) and in some forms of orthostatic hypotension (37) – also improved by carbidopa (38) or cyproheptadine (39) – add credibility to this concept.

Further physiological studies of the part played by DA within this system are needed together with therapeutical trials. The recent development of orally active analogues with natriuretic action which can release DA specifically in the kidney (40) is a particularly hopeful pharmacotherapeutical lead in this field.

Summary

A possible role of DA as an endogenous natriuretic hormone was reviewed under three aspects: (1) its excretion in response to saline vs. albumin-induced volume expansion, (2) the origin of urinary free DA and (3) urinary free DA excretion in an unexplained salt-retaining condition, idiopathic edema. We have shown (1) that the urinary free DA excretion increase in response to saline is specifically related to salt and does not occur when the same degree of volume expansion is induced by albumin, (2) that this increase in free DA originates in the kidney and (3) that idiopathic edema patients excrete less free DA than control subjects.

It is proposed that free DA originating in the kidney is a rather salt than volume-dependent endogenous natriuretic factor. Its deficiency may contribute to excessive sodium retention in idiopathic edema. However, DA is probably not the single natriuretic hormone but a part of a natriuretic system, components of which are other renal vasodilating and natriuretic substances such as prostaglandins and kallikrein-bradykinin.

References

1 *Levinsky, N.:* Natriuretic hormones; in *Levine and Luft* Advances in metabolic disorders, pp. 37–71 (Academic Press, New York 1974).

2 *Adetuybi, A. and Mills, I.H.:* Relation between urinary kallikrein and renal function, hypertension, and excretion of sodium and water in man. Lancet *ii:* 203–207 (1972).

3 *Attallah, A.A. and Lee, J.B.:* Radioimmunoassay of prostaglandin A. Intrarenal PG A_2 as a factor mediating saline induced natriuresis. Circulation Res. *33:* 696–703 (1973).

4 *Goldberg, L.I.:* Dopamine – clinical uses of an endogenous catecholamine. New Engl. J. Med. *291:* 707–710 (1974).

5 *Cuche, J.L.; Kuchel, O.; Barbeau, A.; Boucher, R., and Genest, J.:* Relationship between the adrenergic nervous system and renin during adaptation to upright posture. A possible role for 3,4-dihydroxyphenethylamine (dopamine). Clin. Sci. *43:* 481–489 (1972).

6 *Alexander, R.W.; Gill, J.R., jr.; Yamabe, H.; Lovenberg, W., and Keiser, H.R.:* Effects of dietary sodium and of acute saline infusion on the interrelationship between dopamine excretion and adrenergic reactivity in man. J. clin. Invest. *54:* 194–200 (1974).

7 *Ball, S.G. and Lee, M.R.:* Increased urinary dopamine in salt loaded rats. Clin. Sci. molec. Med. *52:* 20P–21P (1977).

8 *Ball, S.G. and Lee, M.R.:* The effect of carbidopa administration on urinary sodium excretion in man. Is dopamine an intrarenal natriuretic hormone? Br. J. clin. Pharmac. *4:* 115–119 (1977).

9 *Hardaker, W.T., jr. and Wechsler, A.S.:* Redistribution of renal intracortical blood flow during dopamine infusion in dogs. Circulation Res. *33:* 437–444 (1973).

10 *Davis, B.B.; Walter, M.J., and Murdaugh, H.V.:* The mechanism of increase in sodium excretion following dopamine infusion. Proc. Soc. exp. Biol. Med. *129:* 210–213 (1968).

11 *Vlachoyannis, J.; Weismüller, G., and Schoeppe, W.:* Effects of dopamine on kidney function and on the adenyl cyclase phosphodiesterase system in man. Eur. J. clin. Invest. *6:* 131–137 (1976).

12 *Gill, J.R., jr. and Casper, A.G.T.:* Depression of proximal tubular sodium reabsorption in the dog in response to renal beta-adrenergic stimulation by isoproterenol. J. clin. Invest. *50:* 112–118 (1971).

13 *Faucheux, B.; Buu, N.T., and Kuchel, O.:* Effects of saline and albumin on plasma and urinary catecholamines in dogs. Am. J. Physiol. *232:* F123–F127 (1977).

14 *Bell, C. and Land, W.J.:* Neural dopaminergic vasodilator control in the kidney. Nature *246:* 27–29 (1973).

15 *Buu, N.T. and Kuchel, O.:* A new method for the hydrolysis of conjugated catecholamines. J. Lab. clin. Med. *90:* 680–685 (1977).

16 *Buu, N.T.; Unger, T., and Kuchel, O.:* Conjugated catecholamines in doca-saline hypertensive rats. Clin. Res. *26:* 3 (1978).

17 *Unger, T.; Buu, N.T., and Kuchel, O.:* Renal handling of free and conjugated catecholamines following surgical stress in the dog. Am. J. Physiol. (1978).

18 *Kuchel, O.; Cuche, J.L.; Hamet, P.; Buu, N.T.; Nowaczynski, W.; Boucher, R., and Genest, J.:* Idiopathic edema. New pathogenic and therapeutic aspects. Med. Mod. Can. *31:* 614–624 (1976).

19 *Gill, J.R., jr.; Cox, J.; Delea, C.S., and Bartter, F.C.:* Idiopathic edema. II. Pathogenesis of edema in patients with hypoalbuminemia. Am. J. Med. *52:* 452–456 (1972).

20 *Kuchel, O.; Cuche, J.L.; Buu, N.T.; Guthrie, G.P., jr.; Unger, T.; Nowaczynski, W.; Boucher, R., and Genest, J.:* Catecholamine excretion in 'idiopathic' edema. Decreased dopamine excretion, a pathogenic factor? J. clin. Endocr. Metab. *44:* 639–646 (1977).

21 *Gill, J.R., jr.:* The role of the sympathetic nervous system in the regulation of sodium excretion by the kidney; in *Ganong and Martini* Frontiers in neuroendocrinology, chapter 8, pp. 289–305 (Oxford Univ. Press, New York 1969).

22 *Barnardo, D.E.; Summerskill, W.H.L.; Strong, C.G., and Baldus, W.P.:* Renal function, renin activity, and endogenous vasoactive substances in cirrhosis. Am. J. dig. Dis. *15:* 419–425 (1970).

23 *Edwards, C.R.; Miall, P.A.; Hanker, J.P.; Thormer, M.O.; Al-Dujaili, E.A.S., and Besser, G.M.:* Inhibition of the plasma-aldosterone response to frusemide by bromocriptine. Lancet *i:* 903–905 (1975).

24 *Finlay, G.D.; Whitsett, T.L.; Cucinell, E.A., and Goldberg, L.I.:* Augmentation of sodium and potassium excretion, glomerular filtration rate and renal plasma flow from levodopa. New Engl. J. Med. *284:* 865–870 (1971).

25 *Kuchel, O. and Hamet, P.:* Dopamine, extracellular cyclic AMP and sodium excretion. Eur. J. clin. Invest. *7:* 75–76 (1977).

26 *Marieb, N.J. and Mulrow, P.J.:* Failure to escape. A mechanism in idiopathic edema. J. clin. Invest. *43:* 1279 (1964).

27 *Kuchel, O.; Buu, N.T.; Unger, T.; Nowaczynski, W., and Genest, J.:* The dopamine-aldosterone profiles in hypermineralocorticoid edema and hypertension (in preparation).

28 *Kuchel, O.; Buu, N.T.; Faucheux, G.; Unger, T.; Barbeau, A.; Nowaczynski, W. et Genest, J.:* Le rôle de la dopamine dans la régulation de la tension artérielle et de l'excrétion du sodium. Implications physiopathologiques et cliniques. Un. méd. Can. *106:* 507–518 (1977).

29 *Carrière, S.; Lalumière, G., and De Champlain, J.:* Changes in plasma catecholamines in response to isotonic volume contraction or expansion in man and dogs. Clin. Res. *24:* 684A (1976).

30 *Meek, J.L. and Neff, N.H.:* Biogenic amines and their metabolites as substrates for phenol sulphotransferase (EC 2.8.2.1) of brain and liver. J. Neurochem. *21:* 1–9 (1973).

31 *Patak, R.V.; Mookerjee, B.K.; Bentzel, C.J.; Hysert, P.E.; Babej, M., and Lee, J.B.:* Antagonism of the effects of furosemide by indomethacin in normal and hypertensive man. Prostaglandins *10:* 649–659 (1975).

32 *Croxatto, H.R.; Albertini, R.; Arriagada, R.; Roblero, J.; Rojas, M., and Rosas, R.:* Renal artery kallikrein in normotensive and hypertensive rats during enhanced excretion of water and electrolytes. Clin. Sci. molec. Med. *51:* 259s–261s (1976).

33 *Benedek-Jaszmann, J.L. and Hearn-Sturtevant, M.D.:* Premenstrual tension and functional infertility. Aetiology and treatment. Lancet *i:* 1095–1098 (1976).

34 *Stumpe, K.O.; Kolloch, R.; Higuchi, M.; Krück, F., and Vetter, H.:* Hyperprolactinaemia and antihypertensive effect of bromocriptine in essential hypertension. Lancet *ii:* 211–214 (1977).

35 *Tjandramaga, T.B.; Anton, A.H., and Goldberg, L.I.:* Preliminary investigation of levodopa as adjunctive therapy for hypertension; in *Onesti, Kim and Moyer* Hypertension, mechanisms and management, pp. 421–427 (Grune & Stratton, New York 1973).

36 *Gill, J.R., jr.; Fröhlich, J.C.; Bowden, R.E.; Taylor, A.A.; Keiser, H.R.; Seyberth, H.W.; Oates, J.A., and Bartter, F.C.:* Bartter's syndrome. A disorder characterized by high urinary prostaglandins and a dependence of hyperreninemia on prostaglandin synthesis. Am. J. Med. *61:* 43–51 (1976).

37 *Kochar, M.S. and Itskovitz, H.D.:* Hemodynamic and hormonal alterations in idiopathic orthostatic hypotension and response to various modalities of treatment. Clin. Res. *25:* 230A (1977).

38 *Lebel, M.; Kuchel, O.; Barbeau, A.; Hamet, P.; Cuche, J.L.; Nowaczynski, W.; Boucher, R. et Genest, J.:* Excès de dopamine dans l'hypotension orthostatique post-sympathectomie; traitement par un inhibiteur de la dopa-décarboxylase. Un. méd. Can. *102:* 819–825 (1973).

39 *Streeten, D.H.P.; Kerr, L.P.; Kerr, C.B.; Prior, J.C., and Dalakos, T.G.:* Hyper-bradykininism: a new orthostatic syndrome. Lancet *ii:* 1048–1053 (1972).

40 *Kyncl, J.; Hollinger, R.; Warner, R.; Ours, C.W.; Minard, F.N.; Jones, P.H., and Biel, J.H.:* Gamma-glutamyl dopamine, a slow release dopamine prodrug with renal selectivity. Abstr. Am. Soc. Nephrology, Washington 1976, p. 103.

O. Kuchel, MD, Clinical Research Institute, 110 Pine Avenue West, *Montreal, Québec H2W 1R7* (Canada)

Contr. Nephrol., vol. 13, pp. 37–59 (Karger, Basel 1978)

A Concept for the Control of Kidney Production of Erythropoietin Involving Prostaglandins and Cyclic Nucleotides

James W. Fisher, Dennis M. Gross[1], *James E. Foley*[1], *Peter Kim Nelson, George M. Rodgers, William J. George and William Jubiz*

Laboratory of Hematopharmacology, Department of Pharmacology, School of Medicine, Tulane University Medical Center, New Orleans, La., and Veterans Administration Hospital, Department of Medicine, University of Utah, Salt Lake City, Utah

Introduction

Erythropoietin is a hormone secreted by the kidney in response to a variety of physiological and pharmacological stimuli. This hormone is transported from the kidney via the circulation to the bone marrow to affect an increase in the rate of production of red blood cells. Before proposing our model for the role of prostaglandins and cyclic nucleotides in kidney production of erythropoietin, I would like to give a brief history of the discovery of this interesting hormone.

It was in 1906 at the University of Paris that Dr. *Paul Carnot* and Ms. *DeFlandre* discovered that when they injected plasma from donor rabbits which had been bled approximately 25% of their blood volume into recipient rabbits that a marked increase in the peripheral red blood cells of the recipient rabbits occurred (1). Prof. *Carnot* named this substance 'Hemopoietine' (1). Over the next 40 years investigators throughout the world attempted to confirm these findings but were unable to demonstrate this type of erythropoietic activity in the plasma from animals following bleeding. However, in the early 1950s *Erslev* (2) injected large volumes of plasma from bled rabbits, on the order of 5 times that injected by *Carnot,* and did in fact demonstrate a very prompt and marked reticulocytosis. We now know that release of reticulocytes from the bone marrow is one of the very early responses of the marrow to erythropoietin. *Bonsdorff and Jalavisto* (3) later named this hormone *erythropoietin* (Ep) or erythropoietic stimulating factor. At that time it was called a factor because it

[1] Postdoctoral fellows on National Heart, Lung and Blood Institute Hematopharmacology Training Grant (5-T01-HL-05969). This work was also supported by USPHS Grant AM-13211.

was not clear that it was in fact a hormone. In 1950 *Reissmann* (4) of the Air Force Medical Research Laboratories in Texas carried out an interesting series of experiments in which he prepared a parabiotic rat series. Parabiosis is an operation in which the circulation of the lateral skin and muscles of two rats are sutured together which enables the circulation of the two animals to interchange. In this experiment *Reissmann* exposed one partner to an atmosphere of reduced oxygen tension and the second animal was allowed to breathe normal air. Interestingly, erythropoiesis was stimulated in the bone marrow of both animals. The pO_2 levels in animals breathing a reduced oxygen atmosphere was decreased whereas the pO_2 of animals breathing normal air was within normal limits. From this data *Reissmann* proposed that a hormone was being produced by the animal breathing the atmosphere of reduced oxygen pressure which passed across the parabiosis into the recipient animal to stimulate erythropoiesis. Of more clinical interest was the finding of *Stohlman et al.* (5), of a secondary erythrocytosis in a patient with a patent ductus. In that the arterial blood below the level of the diaphragm was markedly hypoxic in this patient, they suggested that the hormone Ep must be produced below the level of the diaphragm.

The clinical finding of an erythrocytosis associated with a patent ductus led *Stohlman* to search for the presence of an erythropoietic factor in the blood of this patient and he eventually was able to demonstrate that this plasma produced a reticulocytosis in recipient animals. However, the assay methods were not particularly good at that time. Quantitative assay methods for Ep were not developed until the middle 1950s when *Leon Jacobson* and his colleagues at the University of Chicago developed a fasted rat assay for Ep which involved radioactive iron incorporation into newly formed red cells (6).

In a search for the site of production of Ep, *Jacobson* and his colleagues systematically removed endocrine organs such as the pituitary, liver (90%), testicles, ovaries, adrenals and the stomach but found that all of the endocrine-deficient animals responded to bleeding and hypoxia to produce Ep in a similar manner to that of the sham-operated animals (7). However, two young medical students at that time, *Walter Fried* and *Louis Plzak,* carried out studies in *Jacobson*'s laboratory where they removed the kidneys and found that the bilaterally nephrectomized animal no longer responded to bleeding or hypoxia by producing Ep (8). This led a number of investigators to focus on the kidney as the site of production of Ep. These studies led *Fisher*'s laboratory (9, 11) and that of *Kuratowska* in Warsaw, Poland (10) to carry out more direct studies on the kidney. We were able to trigger production of Ep in the isolated perfused dog kidney with blood containing cobalt (9), an ion known to increase production of Ep in the kidney. *Kuratowska et al.* (10) carried out hypoxemic perfusion experiments about the same time and produced significant quantities of Ep using the isolated perfused rabbit kidney. We also demonstrated increased Ep production in the isolated hypoxemic perfused dog kidney (11).

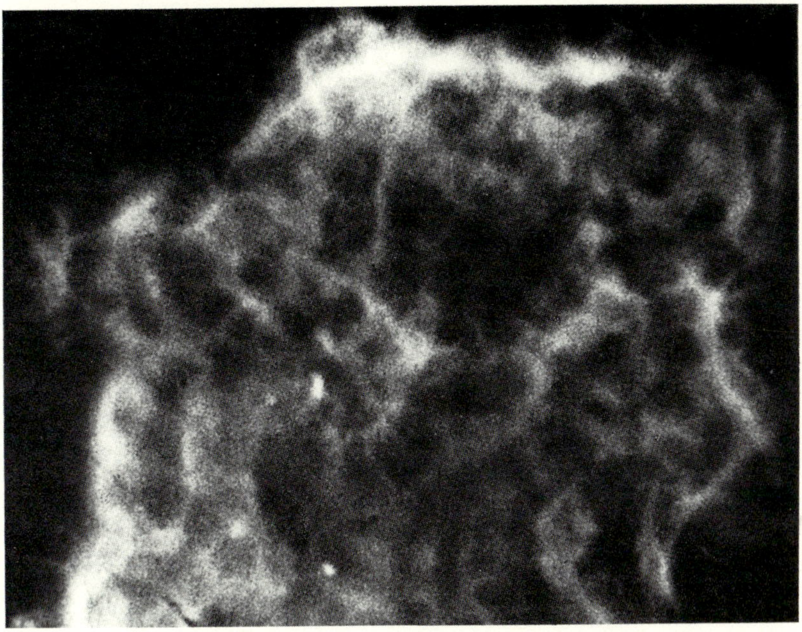

Fig. 1. High power fluorescence photomicrograph of a posthypoxic dog kidney section treated with anti-erythropoietin serum. × 440. From *Busuttil et al.* (16).

The site of production of Ep in the kidney has been investigated for many years. The juxtaglomerular cells (12–14) and glomerular tuft (15–18) have been suggested as sites of production of Ep. *Burlington et al.* (17) produced Ep in a goat renal glomerular tissue culture preparation. *Busuttil et al.* (15, 16) using the indirect fluorescent antibody technique, demonstrated fluorescence in the glomerular tuft of the anemic human (15) and hypoxic dog kidney (16). The specificity of this fluorescence was proven by the ability of highly purified Ep to block the fluorescent staining in the glomeruli. Further studies utilizing electron microscopy (18) suggest that Ep may be localized in the foot processes of the epithelial cells of the glomerular tuft. A photomicrograph of a glomerulus from the kidney of an anemic human subject with positive fluorescent staining for Ep is shown in figure 1.

Role of Prostaglandins in Ep Production

Over the past several years we have been interested in learning whether the prostaglandins (PGs) might mediate kidney generation of Ep. The reason for our identifying the PGs in this system was the fact that several years ago *McGiff et al.* (19) found that renal artery constriction in dogs produced a marked increase in PGE levels in the renal venous effluent. Another reason we became interested in PGs was the fact that *Rodgers et al.* (20) and *George et al.* (21) in our laboratory showed that some of the erythropoietic stimuli activate renal adenylate cyclase, thus causing a rise in cyclic adenosine monophosphate (cAMP) which is probably the first step in the generation of Ep. Cobalt and hypoxia both have been demonstrated to increase cAMP prior to increasing erythrogenin and Ep production (20, 21). The E-type PGs have also been demonstrated to increase tissue levels of cAMP in the liver, thyroid, luteal cell, platelets and lipocyte (22, 23). Thyroid-stimulating hormone has also been found to stimulate arachidonic acid release from phosphatidyl inositol in thyroid homogenates (24).

There appear to be four criteria in our opinion for proving PGs to be the mediator of a biological process.

(1) The PGs should be demonstrated to exert direct effects on the biological process being mediated – in this case kidney production of Ep.

(2) The biological response being mediated should be affected by drugs which inhibit the synthesis of PGs. For these studies the effects of indomethacin on kidney production of Ep following ischemia and hypoxia were used.

(3) Administering PG precursors such as arachidonic acid or the endoperoxides should enhance the biological response postulated to be mediated by the PGs. For these studies arachidonic acid perfusion of the isolated kidney and the effects of arachidonic acid and the endoperoxides on erythropoiesis in polycythemic mice were studied.

(4) A PG receptor antagonist should block the biological response mediated. At the present time we do not have a good PG receptor antagonist but hopefully in the future we will be able to assess this parameter in our studies.

Direct Effects of PG on Erythropoiesis

Schooley and Mahlman (25, 26) have reported the stimulating effect of PGs on erythropoiesis in plethoric mice. These workers demonstrated that PGE_1 and PGE_2 stimulated erythropoiesis in plethoric mice whereas $PGF_{2\alpha}$ was inactive. Figure 2 shows the results of our studies with PGE_2, the naturally occurring E-type PG in the kidney, on radioactive iron incorporation into red cells of exhypoxic polycythemic mice (27). Note the significant increase in radioactive iron incorporation at dosages of 100, 200, 400 and 800 μg/kg compared with 0.2 units Ep. The high dosage required may be due to the rapid inactivation of PGE_2 by the lungs and other tissues and to the relative insensitivity of the polycythemic mice to low levels of Ep. Figure 3 demonstrates that PGA_2, also a

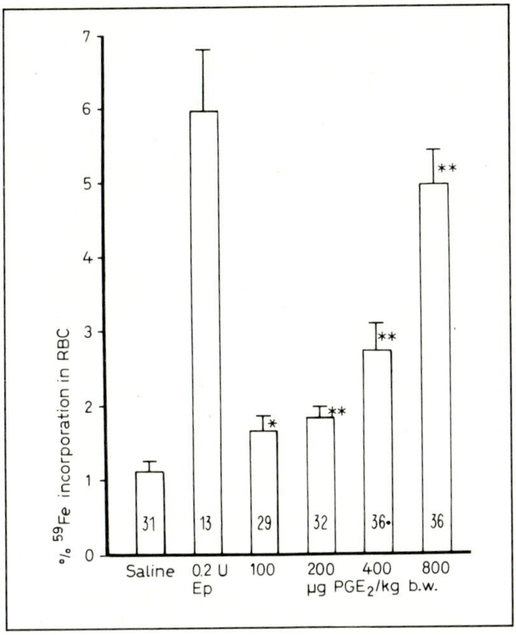

Fig. 2. Percentage of ^{59}Fe incorporation in RBC of exhypoxic polycythemic mice injected with PGE_2. Each bar is the mean ± SE; at the bottom of each bar is the number of mice. Significantly different from saline; * $p < 0.05$, ** $p < 0.01$. From *Gross et al.* (27).

vasodilator, produced a significant increase in iron incorporation in polycythemic mice at dosages of 200, 400 and 800 $\mu g/kg$ body weight.

Figure 4 shows that $PGF_{2\alpha}$, another renal PG, was completely devoid of erythropoietic activity in polycythemic mice.

A requisite for the demonstration that stimuli of erythropoiesis are Ep-dependent events would be the specific blockade of this effect utilizing antiserum to Ep. Figure 5 demonstrates the inhibitory effects of anti-Ep serum when injected in polycythemic mice concurrently at separate sites with 0.2 units human urinary Ep or 400 and 800 $\mu g/kg$ PGE_2. The erythropoietic response elicited by 0.2 units Ep was completely blocked by anti-Ep as was the response elicited by both doses of PGE_2. Along the same lines, PGA_2 was injected concomitantly with anti-Ep serum. Again the erythropoietic response to 0.2 units of Ep was completely abolished and the erythropoietic effects of 400 and 800 $\mu g/kg$ PGA_2 were significantly reduced but not completely abolished. Hence, the blockade of the erythropoietic effects of PGA_2 and PGE_2 by anti-Ep indicates that it is an intrarenal Ep-mediated event, probably a direct action of PGE_2 upon Ep production.

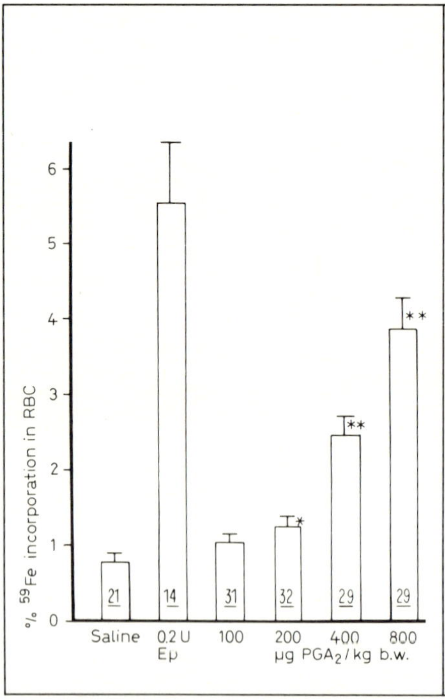

3

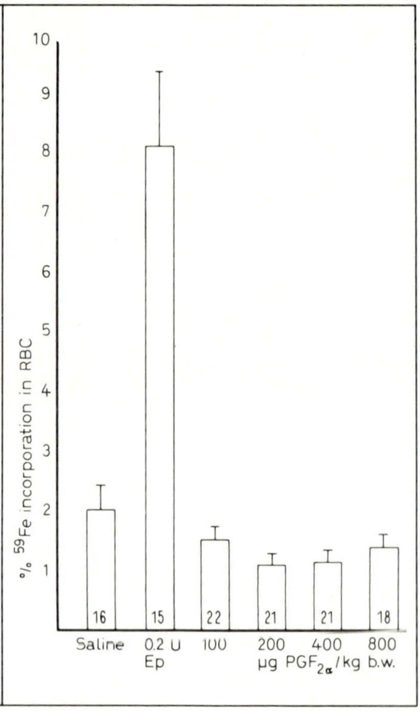

4

Fig. 3. Percentage of ^{59}Fe incorporation in RBC of exhypoxic polycythemic mice injected with PGA_2. Each bar is the mean ± SE; at the bottom of each bar is the number of mice. Significantly different from saline; * $p < 0.02$, ** $p < 0.001$. From *Gross et al.* (27).

Fig. 4. Percentage of ^{59}Fe incorporation in RBC of exhypoxic polycythemic mice injected with $PGF_{2\alpha}$. Each bar is the mean ± SE; at the bottom of each bar is the number of mice. No treatment group mean differed significantly ($p < 0.05$) from the saline control. From *Gross et al.* (27).

In order to study the *in vivo* effects of the primary renal prostaglandin PGE_2 on erythropoiesis, a synthetic long-acting analog of PGE_2, 16,16-dimethyl PGE_2 was studied in exhypoxic polycythemic mice. As indicated in figure 6, 16,16-dimethyl PGE_2 produced a significant ($p < 0.05$) increase in radioiron incorporation in red cells of polycythemic mice. Note in figure 6 that 16,16-dimethyl PGE_2 when compared with PGE_2 (fig. 2) and PGA_2 (fig. 3) produced the most marked increase in radioactive iron incorporation in newly formed red cells in exhypoxic polycythemic mice of any of the PGs studied thus far. Studies are under way to determine the effects of this analog of PGE_2 on total circulating red cell mass in mice following long-term injections and on Ep production utilizing the isolated perfused rabbit kidney.

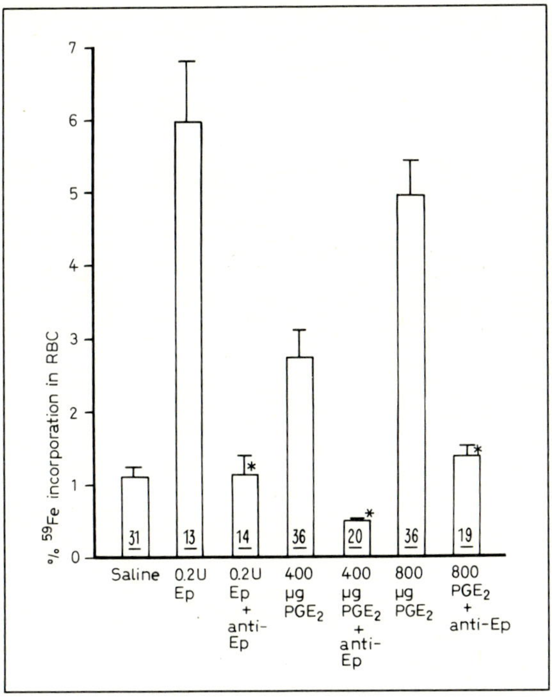

Fig. 5. Inhibition of the erythropoietic effects of PGE_2 by anti-Ep in exhypoxic polycythemic mice. Each bar is the mean ± SE; at the bottom of each bar is the number of mice. Significantly different from control group: * $p < 0.001$. From *Gross et al.* (27).

Our further studies of the direct effects of prostaglandins on erythropoietin production were carried out in the isolated perfused dog kidney.

Figure 7 diagrammatically depicts the isolated perfused kidney system in which posthypoxic dog kidneys were perfused for 5 h. Using this system we decided to study the effects of the primary renal prostaglandin PGE_2 on kidney production of Ep to determine whether the PGs exert a direct effect on kidney production of Ep (27).

Figure 8 presents the results of the effects of PGE_2 infusion on Ep production in the isolated, posthypoxic perfused dog kidneys. Posthypoxic kidneys were perfused for 5 h with blood from nonhypoxic donors. The addition into the perfusion system of PGE_2 at a concentration of 0.1 μg/ml (2.8×10^{-7} *M*) was found in these experiments to be necessary for the stimulation of Ep production in the isolated perfused kidney. A decrease of 31 ± 14.3 mm Hg in renal perfusion pressure was observed within 5 min after the introduction of the

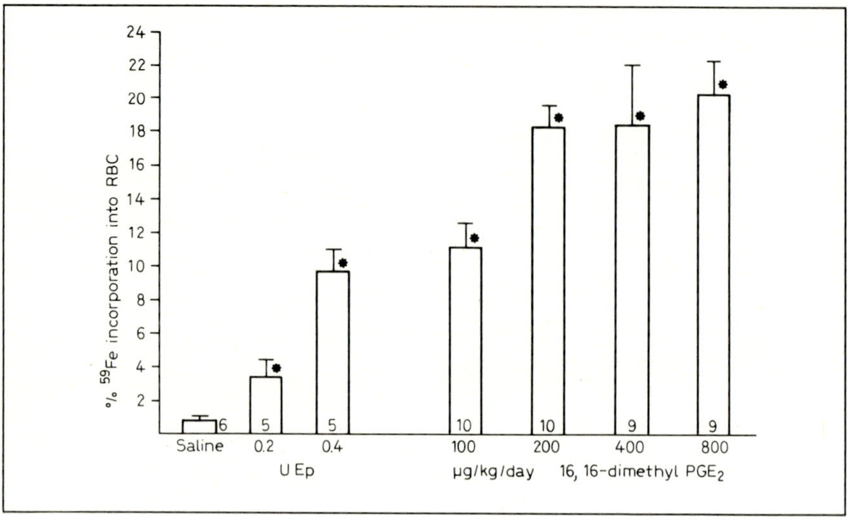

Fig. 6. Percentage of ^{59}Fe incorporation in RBC of exhypoxic polycythemic mice injected with 16,16-dimethyl prostaglandin E_2. Each bar is the mean ± SE; at the bottom of each bar is the number of mice. * Significantly different from saline ($p < 0.05$).

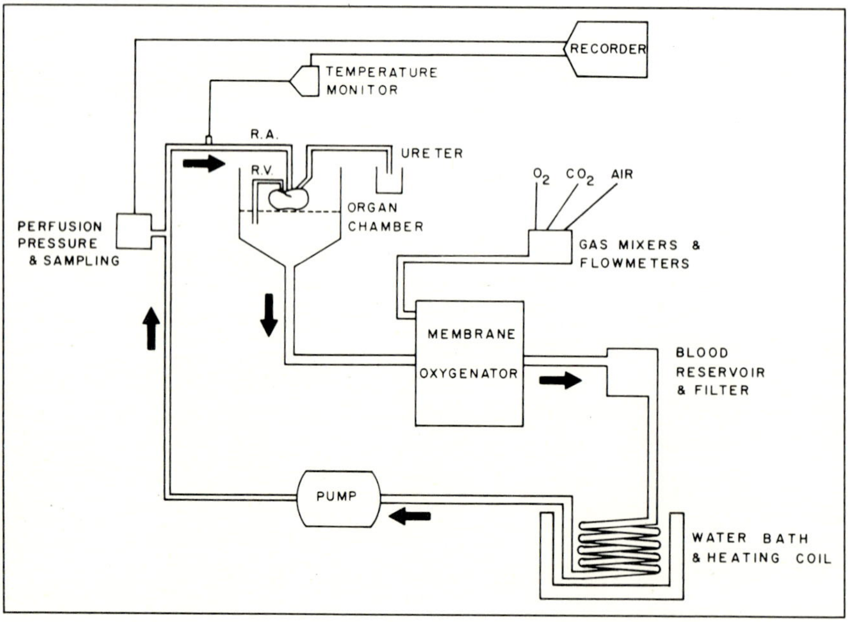

Fig. 7. Isolated kidney perfusion system. Arrows indicate the direction of perfusate flow. From *Gross et al.* (27).

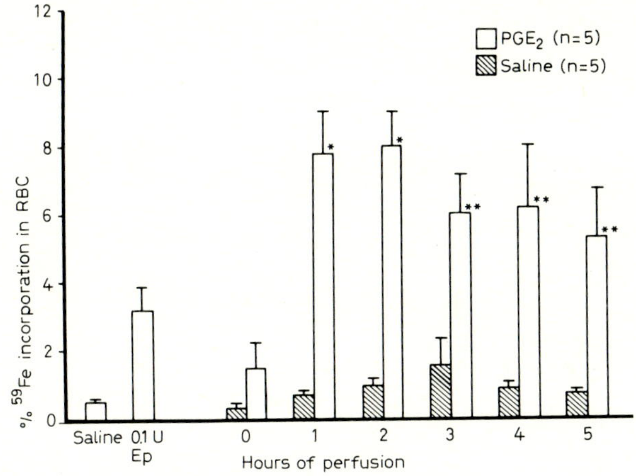

Fig. 8. Effects of PGE_2 on Ep production in the programmed isolated perfused dog kidney. Each bar is the mean ± SE. Significantly different from the respective saline control infusions: * $p < 0.001$, ** $p < 0.05$. From *Gross et al.* (27).

bolus dose of PGE_2 into the perfusion circuit. Following the bolus dose, an infusion was begun at a rate of 0.5 μg/min for the 5-hour experimental period. ^{59}Fe incorporation into red blood cells (RBC) of polycythemic mice injected with perfusates from PGE_2-infused kidneys was found to be significantly increased when compared with that of either the zero hour pre-PGE_2 Ep control levels or the Ep titers in saline-treated control kidneys at the same time intervals. There was no significant difference in saline-infused kidneys at the zero hour Ep titers or at the other time intervals. Samples of the 5-hour PGE_2 perfusates were incubated with anti-Ep and goat anti-rabbit γ-globulin as previously described (27) and were found to block the erythropoietic activity in the perfusates when tested in the exhypoxic polycythemic mouse bioassay.

Inhibition of Ep Production by Cyclo-Oxygenase Inhibitors

We demonstrated several years ago that reduction in renal blood flow to 30–50% of normal renal blood flow in the dog produced a significant increase in Ep production (28). This led us to undertake work on the mechanism by which reduction in renal blood flow may enhance production of Ep (40).

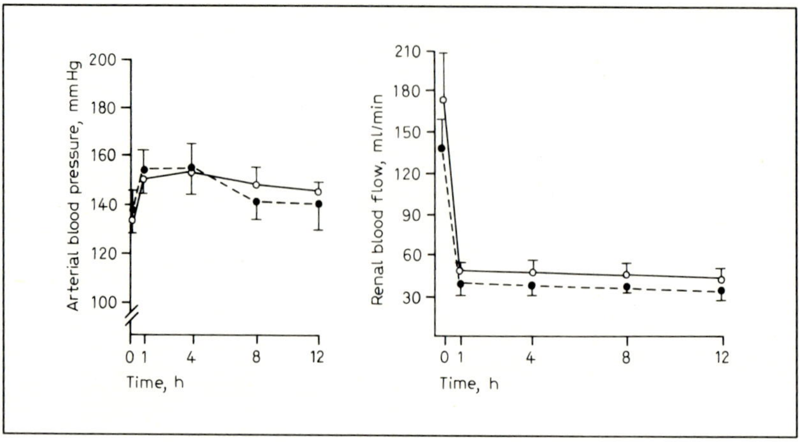

Fig. 9. Arterial blood pressure and renal blood flow in dogs following renal artery constriction (○; n = 6) and indomethacin treatment (•; n = 6). Values are the mean ± 1 SE. From *Gross et al.* (40).

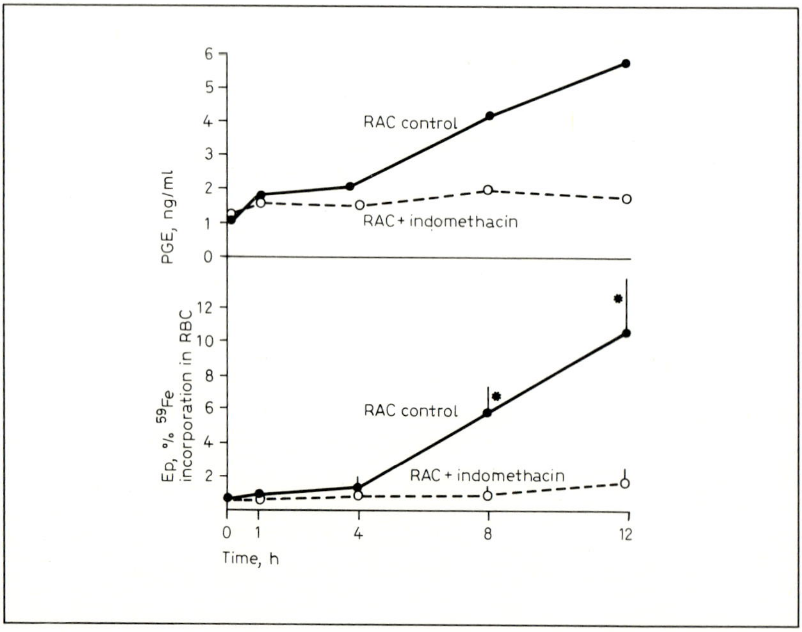

Fig. 10. Effects of indomethacin (I) on serum PGE levels and plasma Ep titers of dogs (n = 6 in each group) following renal artery constriction (RAC) to 30% of normal flow. Values are the mean ± 1 SE. ⊛ Significantly different from RAC + I ($0.1 > p > 0.05$). * Significantly different from RAC + I ($p < 0.05$). From *Gross et al.* (40).

Figure 9 illustrates the arterial pressure and renal blood flow in dogs before and during the first 12 h after renal artery constriction. This data indicates that during the first 12 h after renal artery constriction a significant increase in blood pressure occurs. Under these same conditions, dogs pretreated with indomethacin and renal arteries constricted followed the same pattern of increased arterial pressure as with the renal artery constriction control.

As indicated in figure 10 the mean ^{59}Fe incorporation in RBC of polycythemic mice injected with plasma from 6 dogs at 1, 4, 8 and 12 h following renal artery constriction to 30% of normal flow, showed significantly higher titers of Ep as early as 8 h following the placement of the constriction. In order to determine whether the level of renal artery constriction used in our studies to increase Ep production was sufficient to release renal PGs, renal venous serum PGE levels were determined by radioimmunoassay by *William Jubiz* in Salt Lake City in dogs before and after renal artery constriction and indomethacin pretreatment.

Indomethacin completely blocked the rise in Ep production following renal artery constriction. Note also in figure 10 the changes in PGE levels in the renal

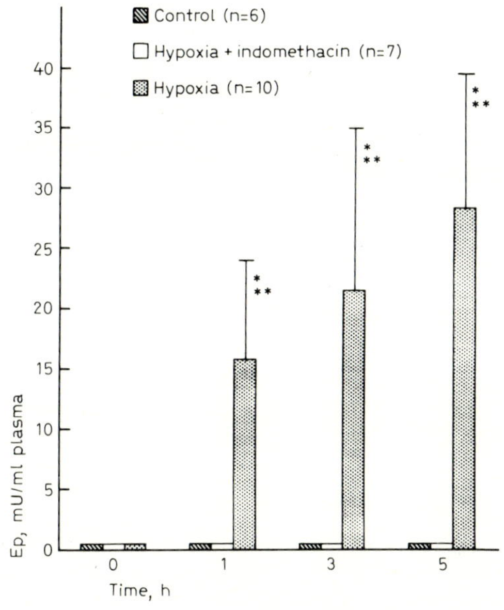

Fig. 11. Effects of hypoxic hypoxia on Ep production.

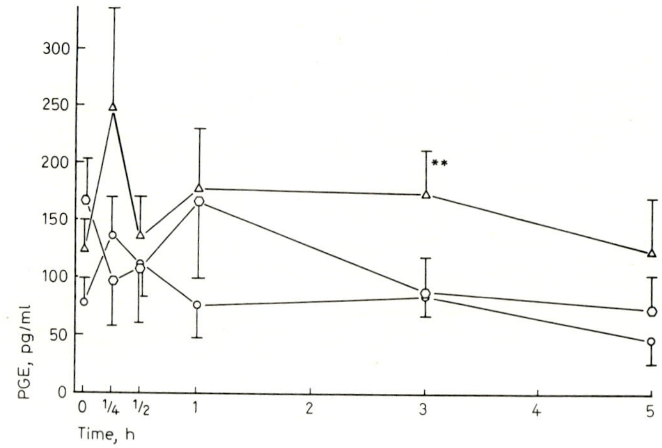

Fig. 12. Effects of hypoxic hypoxia and indomethacin pretreatment on renal venous plasma levels of PGE. O = Control, △ = hypoxic hypoxia, ○ = hypoxic hypoxia + indomethacin.

venous effluent during the course of the 12-hour renal artery constriction. A detectable increase in PGE levels was seen as early as 4 h and a marked increase at 8 h with a highly significant increase in PGE levels after 12 h in the renal venous blood during renal artery constriction to 30% of normal renal blood flow. Again indomethacin completely blocked this rise in PGE levels in renal venous blood. The dosage of indomethacin used *in vivo* was 5 mg/kg orally, 18 and 2 h prior to renal artery constriction.

Further studies on the ability of indomethacin to block Ep production following exposure to erythropoietic stimuli were undertaken utilizing hypoxic hypoxia as the erythropoietic stimulus (32). Female mongrel dogs were allowed to breathe an atmosphere of 8% oxygen and 92% nitrogen for a 5-hour period and Ep, PGE, blood pressure, renal blood flow and renal vascular resistance were measured. Note in figure 11 that hypoxia alone produced a significant increase in Ep levels in plasma as early as 3 h of hypoxic exposure. PGE levels in the renal venous effluent (fig. 12) was increased as early as 15 min and was maintained elevated throughout the hypoxic exposure. Note that indomethacin completely blocked the rise in Ep levels in plasma and the PGE levels in the renal venous effluent following exposure to hypoxia. These data further support the hypothesis that kidney Ep production is related to PGs elaboration.

Effects of PG Precursors on Kidney Production of Ep

As noted previously although the endoperoxides and renal PGs are probably synthesized continuously, it is unlikely that the PGs can act as circulating hormones because of their rapid degradation and target specificity. PGE_2 is a major renal PG and is essential for the regulation of blood distribution between the kidney cortex and medulla and the maintenance of resting blood flow. The finding that PGs are synthesized primarily in the renal medulla, with Ep synthesis probably occurring at cortical sites, pose an interesting problem. However, it is clear that, as *Larsson* (29) has found with mass fragmentography techniques, a significant amount of PGE is synthesized in the kidney cortex. It is therefore possible that the renal cortex is as effective as the medulla in PG biosynthesis. The lower levels of PGs in the cortex could be a consequence of increased prostaglandin metabolism by the enzyme PG dehydrogenase which has been shown to be located primarily within the renal cortex (30). The effects of drugs which inhibit PG synthesis on Ep production establishes the second requirement for PG-mediated erythropoiesis.

Figure 13 is a schematic for the synthesis of PGs and gives the rationale for our studies on the erythropoietic effects of arachidonic acid and the endoperoxides.

As noted in figure 13, arachidonic acid is one of the precursors for the synthesis of PGs and is converted through the cyclo-oxygenase system, the so-called 'PG synthetase', to the endoperoxides PGG_2 and PGH_2. Our previous work has shown that PG synthetase inhibitor drugs such as indomethacin will block Ep production probably by acting on the cyclo-oxygenase system as its site of blockade. It occurred to us that, in order to establish that PGs were involved *in vivo* to the production of Ep, arachidonic acid infusion into the isolated perfused kidney should result in an increase in Ep titers in the perfusates. Therefore, studies were carried out to determine whether arachidonic acid infusion in the isolated perfused kidney would in fact raise Ep titers in the perfusates (39).

Figure 14 demonstrates that arachidonic acid infusion into the isolated perfused kidney at a concentration of 80 μg/min for 5 h produced a significant elevation in titers of Ep in their perfusates. This increase in perfusate levels of Ep in response to arachidonic acid infusion was blocked by indomethacin pretreatment (fig. 14). There has been some support for the hypothesis that the endoperoxides might be the activator of the receptor site in stimulating biological processes and that the PGs may simply be active metabolites. Thus, to study this problem further, the ability of two new stable analogs of PG endoperoxides were tested in our laboratory as to their ability to stimulate erythropoiesis.

The analogs U-44069 [(15S(hydroxy-9α,11α-)epoxymethano) prosta-5Z-

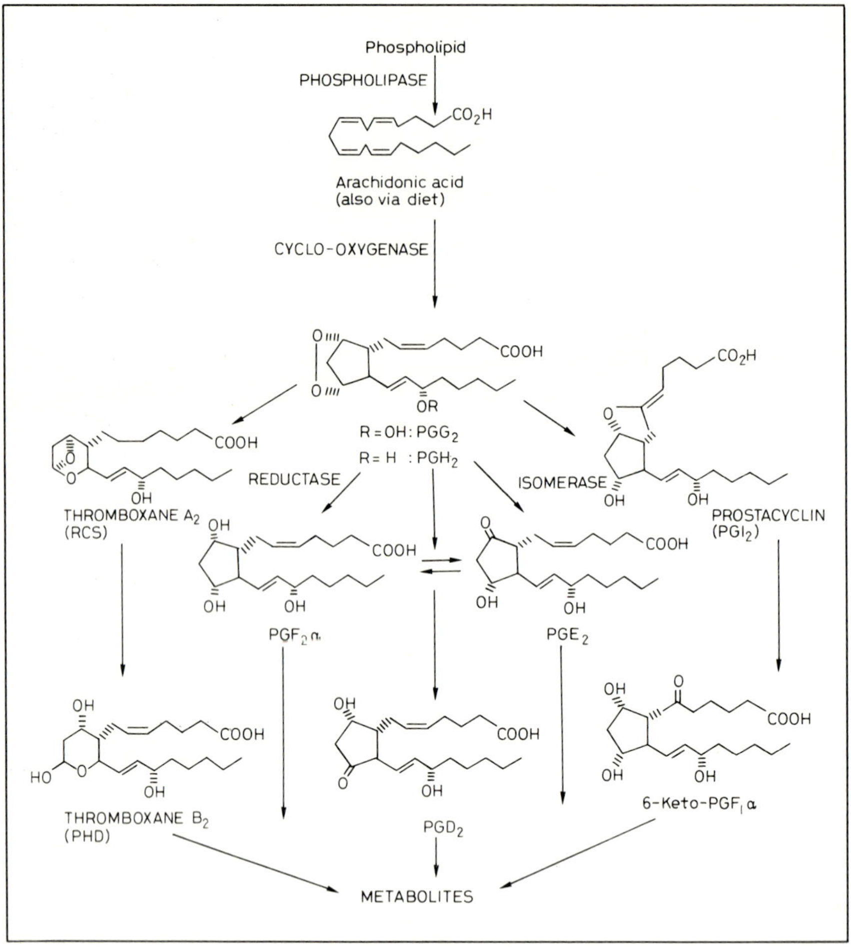

Fig. 13. Schematic pathway for renal prostaglandin synthesis.

13E-dienoic acid] and U-46619 [(15S(hydroxy-11α,9α-)epoxymethano)prosta-5Z-13E-dienoic acid] were tested for their abilities to stimulate radioactive iron incorporation in polycythemic mice (38). These analogs have been tested for their activity in regard to human platelet aggregation and their ability to affect rabbit aorta strips. 11α,9α-Epoxymethano appears to be more potent than the 9α,11α analog in causing platelet aggregation. Their structures are shown in figure 15. Preliminary studies are shown in figure 16 in which polycythemic mice were injected subcutaneously with saline or analog in a dose range of 100–800 μg/kg for 5 consecutive days. Radioiron was administered on day 6

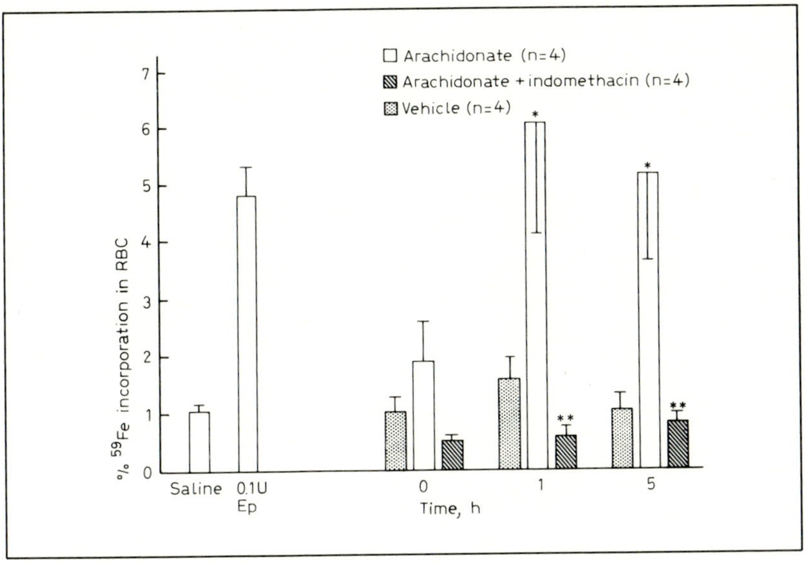

Fig. 14. Effects of arachidonic acid infusion on Ep production in the programmed isolated perfused dog kidney and indomethacin blockade. Each bar is the mean ± SEM and the number of experiments are listed at the base of each bar. Significantly different from the respective saline control: * $p < 0.05$.

and 2 days later the mice were exsanguinated and iron incorporation into newly formed RBC determined. The preliminary data shown in figure 16 demonstrates that except at the highest dose of 11α,9α-epoxymethano (400 μg/kg) tested there was no dose-dependent response evident. Dose of 200, 400 and 800 μg/kg of 9α,11α analog were found to be statistically different from the saline control. In addition, these doses of 9α,11α-epoxymethano were also significantly different from the corresponding doses of 11α,9α analog, apparently the opposite of the platelet aggregation effect of the two endoperoxide analogs.

Relationship between PGE Activation of Adenylate Cyclase and EP Production

Kuehl, (22), *Kuehl and Humes* (30) and *Kuehl et al.* (31) have shown that E type PGs stimulate cAMP formation in a dose-related manner. The correlation between this property and the affinity for membranous receptors seem to have been established. The actions of the E-type PGs have been proposed to be expressed largely via cAMP. Our hypothesis is that renal adenylate cyclase is the receptor for PGE activation of Ep production in the kidney. *Rodgers et al.* (33)

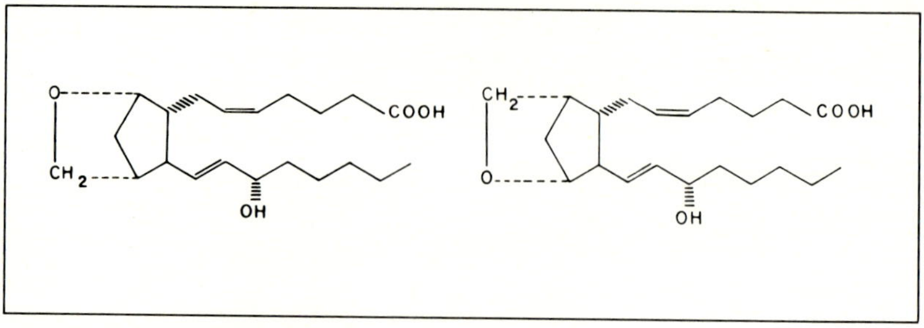

Fig. 15. Chemical structures of endoperoxide analogs.

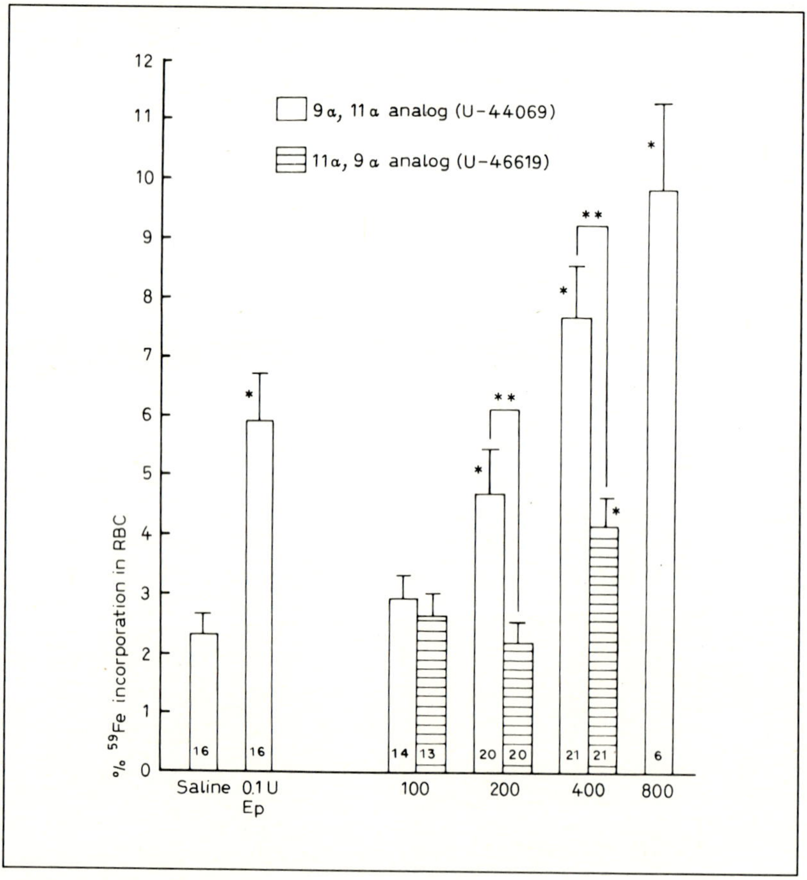

Fig. 16. Percent ^{59}Fe incorporation in RBC of exhypoxic polycythemic mice treated with endoperoxide analogs.

have previously reported that erythropoietic stimuli such as cobalt will produce a marked increase in renal cortical adenylate cyclase activity. *Sutherland et al.* (34) set forth in his early studies the criteria which one must meet in order to justify a claim that a given hormone induces a biological response as the result of an action of cAMP.

The criteria which should be satisfied are: (1) The hormone should be capable of stimulating adenylate cyclase in broken cell preparations in the appropriate tissue. (2) The hormone should be capable of increasing the intracellular levels of cAMP in intact cells. (3) The response to the hormone should be potentiated by the administration of phosphodiesterase inhibitors. (4) The physiological effects of the hormone should be mimicked by the addition of exogenous cAMP.

The effects of PGs on a broken kidney cell adenylate cyclase preparation is under way in our laboratory. We have reported previously that cobalt increased renal cortical adenylate cyclase activity which was related to the mechanism by which it stimulates Ep production (33). *Webster and Beck* (35) recently reported that PGE_2 produced a marked increase in cAMP generation in the kidney cortex and medulla. This increase in cAMP concentration was found to be due to activation of adenylate cyclase.

The second criteria, that the hormone should be able to increase the intracellular levels of cAMP in intact cells, needs to be further established. In that it is postulated that hypoxia activates adenylate cyclase in the kidney by an early release of PGE_2, it is necessary first to determine whether hypoxia increases kidney levels of cAMP. As noted in figure 17, two peaks in plasma Ep occurred in rats during 24 h exposure to hypoxia (0.42 atm). As shown in figure 18 renal cAMP levels were significantly elevated in rat kidneys following exposure to hypoxia (42). As depicted (fig. 18) there are two significant ($p < 0.05$) renal cAMP peaks (42). We have demonstrated previously that PGE_1 increases Ep production and elevates cAMP levels in the isolated perfused dog kidney (41). However, it is necessary to establish that the naturally occurring prostaglandin PGE_2 or one of its intermediates will increase cAMP levels in the isolated perfused kidney.

The third criteria for establishing that a hormone action is due to an adenylate cyclase mediated response is to demonstrate that a phosphodiesterase inhibitor will potentiate the effects of the hormone. As noted in table I, *Rodgers et al.* (33) have demonstrated potentiation of the erythropoietic effects of cobalt by theophylline. Further studies are under way to determine whether the erythropoietic effects of PGE_2 are potentiated by theophylline.

The fourth criteria that the physiological effects of the hormone is mimicked by the addition of exogenous cAMP has already been met in our earlier work (33, 36). As indicated in figure 19, these earlier studies (33, 36) clearly demonstrate that the dibutyryl derivative of cAMP produced a significant

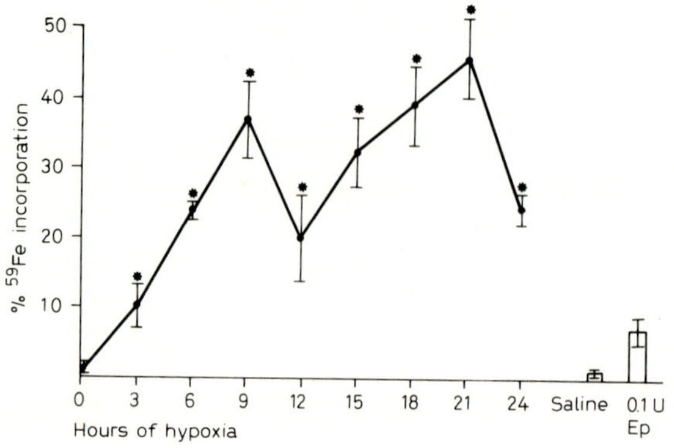

Fig. 17. Plasma Ep levels following exposure to hypobaric hypoxia in the rat. * Significantly different from control ($p < 0.05$). From *George et al.* (42).

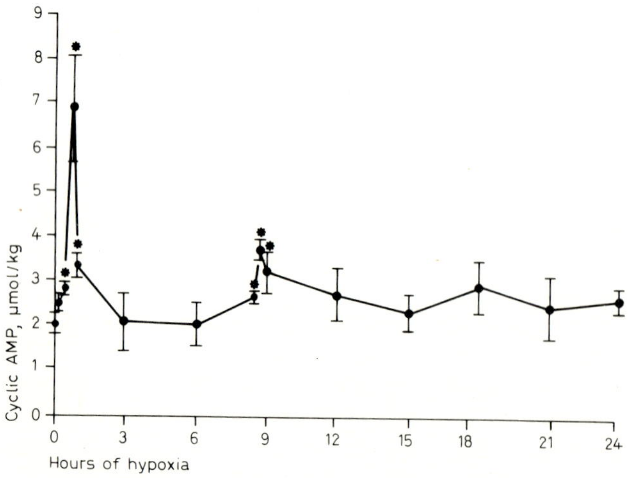

Fig. 18. Renal cAMP levels following exposure to hypobaric hypoxia in the rat. * Significantly different from control ($p < 0.05$). From *George et al.* (42).

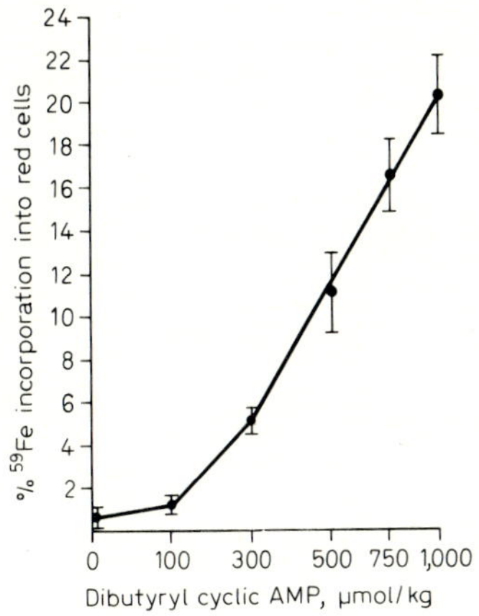

Fig. 19. Dose response relationship of dibutyryl cAMP in the polycythemic mouse assay. From *Rodgers et al.* (20).

Table I. Potentiation of the erythropoietic effects of cobalt by theophylline

Plasma from animals treated with	IRP units of Ep[1]
Saline	0.04 ± 0.007
Theophylline (80 mg/kg)	0.03 ± 0.01
Cobalt (150 µmol/kg)	0.47 ± 0.06
Cobalt (150 µmol/kg) + theophylline (80 mg/kg)	0.90 ± 0.21[2]

[1] Mean of 3 experiments (5 polycythemic mice per experiment).
[2] Significantly different from all controls ($p < 0.05$).
From *Rodgers et al.* (33).

increase in radioactive iron incorporation in red cells of polycythemic mice which was greater than that of 0.2 units Ep. In addition, the antibody to Ep abolished for the most part the erythropoietic response of polycythemic mice to dibutyryl cAMP (20, 25, 37). Also note in table II that dibutyryl cAMP produced a significant increase in hematocrit and red cell mass in normal mice injected for a period of 14 days (36). Thus, it appears that the criteria have already been met which would indicate that exogenous cAMP is capable of stimulating erythropoiesis.

Table II. Effects of dibutyryl cAMP and Ep on the hematocrit and red cell mass in normal mice

Group	Final body weight, g	Hematocrit %	Red cell mass ml/100 g
Control	29.0 ± 1.05	43.25 ± 0.85	2.93 ± 0.20
Dibutyryl cAMP (8 mg, s.c., daily)	28.4 ± 0.24	46.50 ± 0.64*	4.17 ± 0.08*
Erythropoietin (0.6 units, s.c., daily)	29.0 ± 1.30	51.40 ± 1.54*	4.57 ± 0.39*

* Experimental values significantly ($p < 0.05$) different from that of the controls.
From *Rogers et al.* (36).

Conclusion

What we would like to present in conclusion is our hypothesis concerning the possible role of the PGs in Ep production. As indicated in figure 20 we postulate that hypoxia, either ischemic or hypoxic hypoxia, can cause the *de novo* synthesis of PGs, principally PGE_2. The stimulus of hypoxia most likely interacts at the level of cyclo-oxygenase or perhaps at an earlier step such as on phospholipase enhancing the availability of arachidonic acid, the rate-limiting step in the PG cascade. This leads to the formation of the endoperoxides, which we have demonstrated are erythropoietically active, and the formation of PGE_2, the primary renal PG. The PGE_2 probably activates a renal cortical adenylate cyclase to generate intracellular cAMP. The renal cAMP elevation results in the activation of a protein kinase which interacts with erythrogenin (renal erythropoietic factor) or a proerythropoietin to cause an increase in production and/or secretion of Ep.

Summary

Our hypothesis is that PGs released within the kidney play a role in the modulation of kidney production of Ep. PGs release probably at medullary and/or cortical sites following erythropoietic stimuli such as hypoxic hypoxia, anemic hypoxia, and ischemic hypoxia induced by renal artery constriction increases kidney production of Ep. PGs which are released probably activate a renal cortical adenylate cyclase thereby enhancing the production of intracellular cAMP. This initiates the cascade of events resulting in the production and/or secretion of Ep by the kidney. The endoperoxide analogs and PGE_2 have been found to produce a dose-related and Ep-dependent increase in radioiron incorporation into newly formed red blood cells of exhypoxic polycythemic mice. Indomethacin, a potent PG cyclo-oxygenase inhibitor, attenuates Ep production and the appearance of PGE in the renal venous effluent of animals exposed to hypoxic hypoxia and renal artery constriction. Arachidonic acid (C20:4) and PGE_2 infusion into the posthypoxic isolated perfused dog

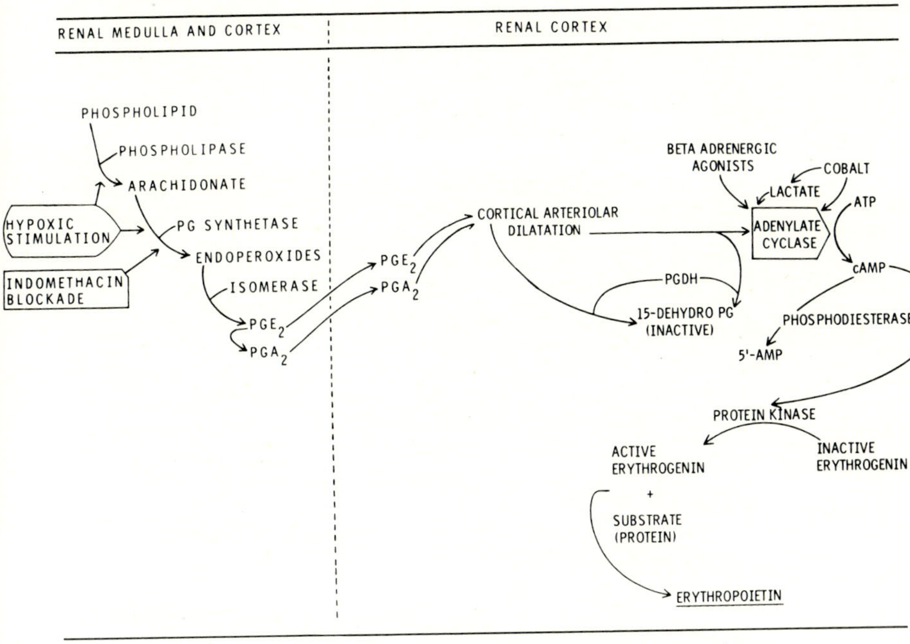

Fig. 20. Schematic hypothesis for PG-induced kidney production of Ep.

kidney produced a significant elevation in Ep titers in the perfusate. The increase in Ep production caused by arachidonate is blocked by indomethacin. It has been previously reported that PGs of the E series stimulate cAMP formation in several tissues. We have found that not only are renal cortical cAMP levels significantly elevated in rats following exposure to hypobaric hypoxia but that dibutyryl cAMP administration produces an increase in hematocrit and circulating red cell mass in normal mice. Our data thus far strongly support the hypothesis that the renal PGs and the cyclic nucleotides are intimately involved in the pharmacologic and/or pathophysiologic control of Ep production. Further work is necessary to determine whether the PGs and cyclic nucleotides are involved in the day-to-day control of Ep production by the mammalian kidney.

References

1 *Carnot, P. et DeFlandre, C.:* Sur l'activité hémapoïetique des différents organes au cours de la régénération du sang. C.r. hebd. Séanc. Acad. Sci., Paris *143:* 432 (1906).

2 *Erslev, A.J.:* Humoral regulation of red cell production. Blood *8:* 349 (1953).

3 *Bonsdorff, E. and Jalavisto, E.:* A humoral mechanism in anoxic erythrocytosis. Acta physiol. scand. *16:* 150 (1948).

4 *Reissmann, K.R.:* Studies on the mechanism of erythropoietic stimulation in parabiotic rats during hypoxia. Blood *5:* 372 (1950).

5 *Stohlman, F., jr.; Rath, C.E., and Rose, J.C.:* Evidence for a humoral regulation of erythropoiesis. Studies on a patient with polycythemia secondary to regional hypoxia. Blood *9:* 721 (1954).

6 *Plzak, L.F.; Fried, W.; Jacobson, L.O., and Bethard, W.F.:* Demonstration of stimulation of erythropoiesis by plasma from anemic rats using ^{59}Fe. J. Lab. clin. Med. *46:* 671 (1955).

7 *Jacobson, L.O.; Goldwasser, E.; Fried, W., and Plzak, L.:* Studies on erythropoiesis. VII. The role of the kidney in the production of erythropoietin. Trans. Ass. Am. Physns *70:* 305 (1957).

8 *Jacobson, L.O.; Goldwasser, E.; Fried, W., and Plzak, L.:* Role of the kidney in erythropoiesis. Nature, Lond. *179:* 633 (1957).

9 *Fisher, J.W. and Birdwell, B.J.:* The production of an erythropoietic factor by the *in situ* perfused kidney. Acta haemat. *26:* 244 (1961).

10 *Kuratowska, Z.; Lewartowski, B., and Michalak, E.:* Studies on the production of erythropoietin by isolated perfused organs. Blood *18:* 527 (1961).

11 *Fisher, J.W. and Langston, J.W.:* The influence of hypoxemia and cobalt on erythropoietin production in the isolated perfused dog kidney. Blood *29:* 114 (1967).

12 *Goldfarb, B. and Tobian, L.:* The interrelationship of hypoxia, erythropoietin, and the renal juxtaglomerular cell. Proc. Soc. exp. Biol. Med. *111:* 510 (1962).

13 *Hirashima, K. and Takaku, F.:* Experimental studies on erythropoietin. II. The relationship between juxtaglomerular cells and erythropoietin. Blood *20:* 1 (1962).

14 *Mitus, W.J.; Toyama, K., and Brauer, M.J.:* Erythrocytosis, juxtaglomerular apparatus (JGA) and erythropoietin in the course of experimental unilateral hydronephrosis in rabbits. Ann. N. Y. Acad. Sci. *149:* 107 (1968).

15 *Busuttil, R.W.; Roh, B.L., and Fisher, J.W.:* The cytological localization of erythropoietin in the human kidney using the fluorescent antibody technique. Proc. Soc. exp. Biol. Med. *137:* 327 (1971).

16 *Busuttil, R.W.; Roh, B.L., and Fisher, J.W.:* Further evidence for the glomerulus as a site of production of erythropoietin in the hypoxic dog kidney. Acta haemat. *47:* 238 (1972).

17 *Burlington, H.; Cronkite, E.P.; Reineke, U., and Zanazni, E.:* Erythropoietic activity produced in cultures of renal glomeruli. Abstr. 14th Annu. Meet. Am. Soc. Hematol., p. 827 (1971).

18 *Busuttil, R.W.; Chen, I.; Yates, R., and Fisher, J.W.:* The cytological localization of erythropoietin in the hypoxic dog kidney using immune peroxidase method and electron microscopy. Abstr. Meet. Am. Soc. Hematol., Atlanta 1974, p. 111.

19 *McGiff, J.C.; Crowshaw, K.; Terragno, N.A.; Lonigro, A.J.; Strand, J.C.; Williamson, M.A.; Lee, J.B., and Ng, K.K.F.:* Prostaglandin-like substances appearing in canine renal venous blood during renal ischemia. Circulation Res. *27:* 765 (1970).

20 *Rodgers, G.M.; George, W.J., and Fisher, J.W.:* Effects of cobalt on kidney cyclic AMP renal erythropoietic factor and erythropoietin production. Proc. Soc. exp. Biol. Med. *140:* 977 (1972).

21 *George, W.J.; Briggs, D.U., and Fisher, J.W.:* Metabolism of erythropoietin; in *Fisher* Kidney hormones, vol. II, *13:* 109 (1977).

22 *Kuehl, F.A.:* Prostaglandins, cyclic nucleotides and cell function. Prostaglandins *5:* 325 (1974).

23 *Gorman, R.:* Prostaglandin endoperoxides. New regulators of cyclic nucleotide. J. cycl. Nucl. Res. *1:* 1 (1975).

24 *Haye, B.; Champion, S., and Jacquemin, C.:* FEBS Lett. *30:* 253 (1973).

25 *Schooley, J.C. and Mahlmann, L.J.:* Stimulation of erythropoiesis in the plethoric

mouse by cyclic-AMP and its inhibition by anti-erythropoietin. Proc. Soc. exp. Biol. Med. *137:* 1289 (1971).

26 *Schooley, J.C. and Mahlmann, L.J.:* Stimulation of erythropoiesis in plethoric mice by prostaglandins and its inhibition by antierythropoietin. Proc. Soc. exp. Biol. Med. *138:* 523 (1971).

27 *Gross, D.M.; Brookins, J.; Fink, G.D., and Fisher, J.W.:* Effects of prostaglandins A_2, E_2 and $F_{2\alpha}$ on erythropoietin production. J. Pharmac. exp. Ther. *198:* 489 (1976).

28 *Fisher, J.W.; Samuels, A.I., and Langston, J.W.:* Effects of angiotensin and renal artery constriction on erythropoietin production. J. Pharmac. exp. Ther. *157:* 618 (1967).

29 *Larsson, C.:* The renal prostaglandins, an experimental study in the rabbit; thesis, Stockholm (1975).

30 *Kuehl, F.A. and Humes, J.L.:* Direct evidence for a prostaglandin receptor and its application to prostaglandin measurements. Proc. natn. Acad. Sci. USA *69:* 480 (1972).

31 *Kuehl, F.A., jr.; Cirillo, V.J.; Ham, E.A., and Humes, J.L.:* The regulatory role of the prostaglandins on the cyclic 3',5'-AMP system. Adv. Biosci. *9:* 155 (1973).

32 *Gross, D.M.; Jubiz, W., and Fisher, J.W.:* Release of erythropoietin and prostaglandin E by the dog kidney during hypoxic hypoxia and the effects of indomethacin. Fed. Proc. Fed. Am. Socs exp. Biol. *36:* 1052 (1977).

33 *Rodgers, G.M.; Fisher, J.W., and George, W.J.:* The role of renal adenosine 3',5'-monophosphate in the control of erythropoietin production. Am. J. Med. *58:* 31 (1975).

34 *Sutherland, E.W.; Robison, G.A., and Butcher, R.W.:* Some aspects of the biological role of adenosine 3',5'-monophosphate (cyclic AMP). Circulation *37:* 279 (1968).

35 *Webster, S.K. and Beck, N.:* Interaction of prostaglandin E_2 and angiotensin II on cyclic AMP in the kidney and the aorta. Abstract. Fed. Proc. Fed. Am. Socs exp. Biol. *402:* (1977).

36 *Rodgers, G.M.; Fisher, J.W., and George, W.J.:* Increase in hematocrit, hemoglobin and red cell mass in normal mice after treatment with cyclic AMP. Proc. Soc. exp. Biol. Med. *148:* 380 (1975).

37 *Peschle, C.; Rappaport, I.A.; D'Avanzo, A.; Russolillo, S.; Marone, G., and Condorelli, M.:* Renal mechanisms underlying cyclic AMP action on erythropoiesis. Br. J. Haemat. *25:* 393 (1973).

38 *Gross, D.M.; Foley, J.E.; Brookins, J., and Fisher, J.W.:* The effects of endoperoxide analogs and prostaglandin E_2 on erythropoietin production. Pharmacologist *18:* 216 (1976).

39 *Foley, J.E.; Gross, D.M.; Brookins, J., and Fisher, J.W.:* The effects of arachidonic acid on erythropoietin production. Pharmacologist *18:* 216 (1976).

40 *Gross, D.M.; Mujovic, V.M.; Jubiz, W., and Fisher, J.W.:* Enhanced erythropoietin and prostaglandin E production in the dog following renal artery constriction. Proc. Soc. exp. Biol. Med. *151:* 498 (1976).

41 *Paulo, L.G.; Wilkerson, R.D.; Roh, B.L.; George, W.J., and Fisher, J.W.:* Effects of prostaglandin E_1 on erythropoietin production. Proc. Soc. exp. Biol. Med. *142:* 771 (1973).

42 *George, W.J.; Rodgers, G.M.; Briggs, D.W., and Fisher, J.W.:* Role of cyclic AMP in the regulation of erythropoietin production; in *Nakao, Fisher and Takaku* Erythropoiesis. Proc. 4th Int. Conf. Erythropoiesis, pp. 277–291 (University of Tokyo Press, Tokyo 1975).

Dr. *J.W. Fisher,* Laboratory of Hematopharmacology, Department of Pharmacology, School of Medicine, Tulane University Medical Center, *New Orleans, La.* (USA)

Contr. Nephrol., vol. 13, pp. 60–68 (Karger, Basel 1978)

Absolute and Relative Erythropoietin Deficiency in Renal Anemia

K.M. Koch and H.W. Radtke

Department of Nephrology, Center of Internal Medicine, University of Frankfurt, Frankfurt (Main)

Introduction

Even though there is ample proof that a number of mechanisms are involved in the pathogenesis of renal anemia (1, 2), lack of erythropoietin (Ep) is considered to be the most important cause for the development of anemia in chronic renal failure (3, 4). This assessment of the role of Ep deficiency is based on the observation that the most severe degrees of renal anemia occur in bilateral nephrectomized patients (5, 6). Additional support for this concept is derived from the fact that so far only rarely elevated serum erythropoietin (SEp) levels could be demonstrated in azotemic anemic patients (7–17). As in general the degree of anemia correlates with the severity of azotemia (1, 18), it is assumed that decrease of excretory ability of the kidney is accompanied by an equivalent reduction of incretory renal function. No matter how plausible this concept of a primary role of lack of Ep in the pathogenesis of renal anemia appears to be, its validity in the case of nonnephrectomized patients still has to be established. Because of insensitive bioassays applied in most of the studies and the small number of patients investigated, the actual data on SEp levels in chronic renal failure are scarce.

In the present study, the relation between SEp concentration and anemia was investigated by means of a highly sensitive bioassay in a large number of regular hemodialysis patients.

Patients and Methods

A total of 100 (27 females, 73 males) regular hemodialysis patients were investigated. All patients were treated by 3 × 6–10 h/week hemodialysis on 1 m^2 dialyzers with 11.5–13.5 μm Cuprophane membranes for at least 6 months. None required intermittent transfusion therapy. All patients were on oral or intravenous iron substitution and 14 received testosterone propionate 100–250 mg/week intramuscularly.

In addition, measurements of SEp concentration were performed in 13 bilateral nephrectomized patients (3 of them with intermittent transfusion therapy and 8 receiving 250–500 mg testosterone propionate intramuscularly per week), 12 azotemic patients on the day of first dialysis, 9 anemic patients without renal disease and 30 healthy controls. In those patients who had been studied on the day of first dialysis, repeat measurements were performed 3–27 months following the onset of treatment. The dialysis regimen was identical with that of all other hemodialysis patients investigated.

Measurements of SEp concentrations were performed by use of fetal mouse liver cell culture as described by *Dunn et al.* (19) and *Wardle et al.* (20). Hemopoietic livers were harvested from 14-day-old mouse fetuses and suspended in a culture medium consisting of 85% minimal essential medium, 10% fetal calf serum, and 5% bicarbonate buffer in CO_2 atmosphere. Fixed volume aliquots of serial dilution of test serum or standard were added to constant volumes of cell suspension and five replicas were made of every concentration step. After a preincubation period of 20 h at 37 °C, radioiron transferrin was added and the culture incubated for another 4 h for pulse labelling. Subsequently, cellular lysis in distilled water was performed and heme extracted using HCl and Drabkin's solution in butanone. Heme radioactivity was measured in a β-liquid scintillation counter. Rate of radioiron incorporation into heme served as a measure of Ep activity.

In order to be able to express SEp concentration in international units, we calibrated our laboratory standard (commercial sheep plasma standard step III, Connaught, Toronto) by use of the International Standard Preparation B, courteously made available through WHO International Laboratory for Biological Standards, Mill Hill, London (21). As the fetal mouse liver assay is sensitive down to 0.5 mU Ep standard/ml final dilution and the test serum portion of culture medium is 10% at the maximum SEp concentrations, down to 5 mU/ml can be measured accurately. The calculation of Ep titer with 95% confidence limits and testing of significance of regression, linearity and parallelism were performed by analysis of variance (22).

Results

Figure 1 shows the results of 100 regular hemodialysis patients studied. SEp concentration is plotted against the corresponding hematocrit. SEp concentration ranges from 23 to 389 mU/ml with a mean of 135 mU/ml. In patients with hematocrits higher than 40 vol% SEp concentration is approximate to 120 mU/ml. This SEp level is almost identical with the mean of 126 ± (SD) 50 mU/ml found in 30 healthy controls. From the data, two distinctive groups of patients became apparent. In the first group decreasing hematocrits are accompanied by decreasing SEp concentrations, whereas in the second group SEp concentrations are rising with dropping hematocrits. The first group consists of patients with SEp concentrations below 120 mU/ml whereas in the second group all SEp concentrations are above 120 mU/ml. In figure 2 the values of the first group are presented separately. The mean hematocrit is 29 vol%, mean SEp concentration 68 mU/ml. There is a highly significant positive correlation be-

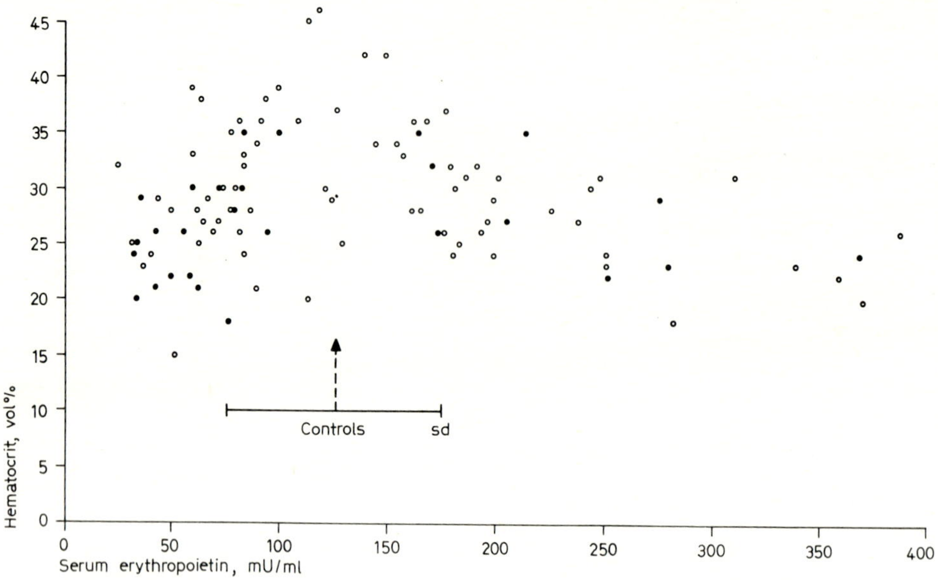

Fig. 1. SEp concentration and hematocrit in 100 regular hemodialysis patients. Mean ± SD of 30 healthy controls is indicated by the dotted line. ○ = Males (n = 73); ● = females (n = 27).

tween SEp levels and hematocrits (r = 0.50, p <0.001). In contrast, the results of the second group (fig. 3) exhibit a highly significant negative correlation (r = −0.63, p <0.001). The mean hematocrit again is 29 vol%, mean SEp concentration 211 mU/ml.

Discussion

The necessity of maintenance hemodialysis treatment is proof of a minimal residual renal excretory function in all patients investigated. In contrast, incretory renal function varies widely as judged from SEp concentrations ranging from very low levels to values three times higher than the mean of normal controls. In more than half of the patients, SEp concentrations are lower than the mean of normal controls, i.e. absolutely decreased. In these patients Ep deficiency appears to play a primary role in the pathogenesis of anemia, as can be derived from the fact that hematocrit and SEp concentrations are positively correlated. Higher SEp levels are accompanied by a lesser degree of anemia. This

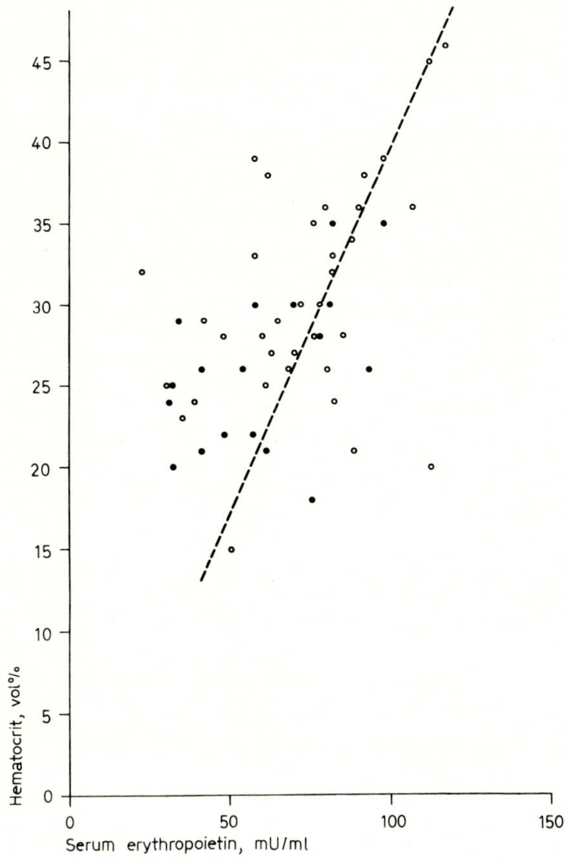

Fig. 2. Correlation between SEp concentration and hematocrit in 53 regular hemodialysis patients with SEp concentrations lower than 120 mU/ml. ○ = Males (n = 35); • = females (n = 18); r = 0.50; p <0.001; Y = –4.9 + 0.45 X.

is the pattern we also found in anephric patients where Ep deficiency is known to be extreme (23, 24). In a group of 13 bilateral nephrectomized subjects with a hematocrit ranging from 15 to 29 vol%, SEp concentrations and hematocrits were significantly positively correlated (r = 0.68, p <0.01). The mean SEp concentration was 114 mU/ml and individual values ranged from 30 to 215 mU/ml.

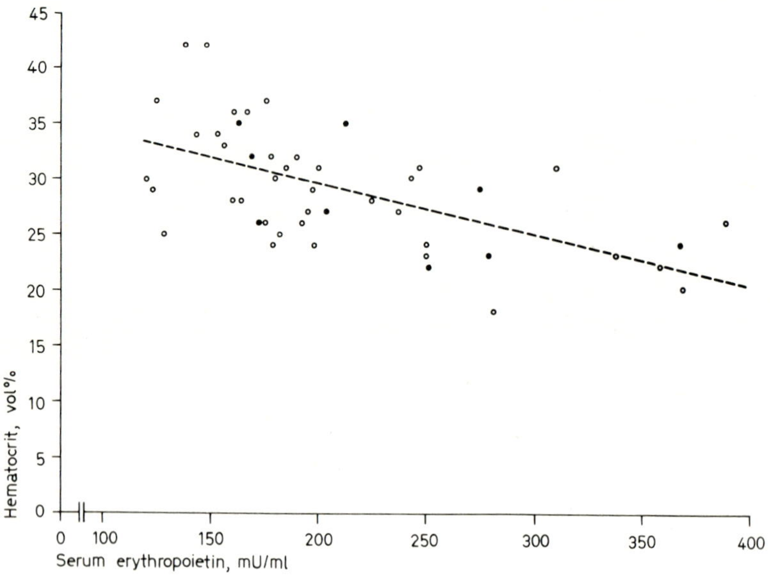

Fig. 3. Correlation between SEp concentrations and hematocrit in 47 regular hemodialysis patients with SEp concentrations higher than 120 mU/ml. ○ = Males (n = 38); • = females (n = 9); r = –0.63; p <0.001; Y = 38.9 –0.047 X.

In the remaining patients with SEp levels above the mean of normal controls, the kidneys obviously are able to respond to anemia by producing increased amounts of Ep. However, when compared to 9 severe anemic patients (mean hematocrit 23 vol%) without renal disease, where SEp levels ranged from 333 to 996 mU/ml, SEp levels in these patients are low. Nevertheless, one has to conclude that in this group of patients other factors, such as uremic toxicity and blood loss, may be primary causes of anemia. This concept is supported by our results in 12 azotemic patients who were studied on the first day of dialysis and 3–27 months following the onset of maintenance hemodialysis. Before treatment, the mean SEp concentration of these patients was 520 ± (SD) 567 mU/ml, while the mean hematocrit was 21 ± (SD) 4.0 vol%.

When measurements were repeated, mean SEp concentration had dropped to 206 ± 94 mU/ml, while mean hematocrit had risen to 31 ± 5.0 vol%. Individual values are shown in figure 4. As can be seen, every patient experienced a decrease in SEp concentration and an increase in hematocrit.

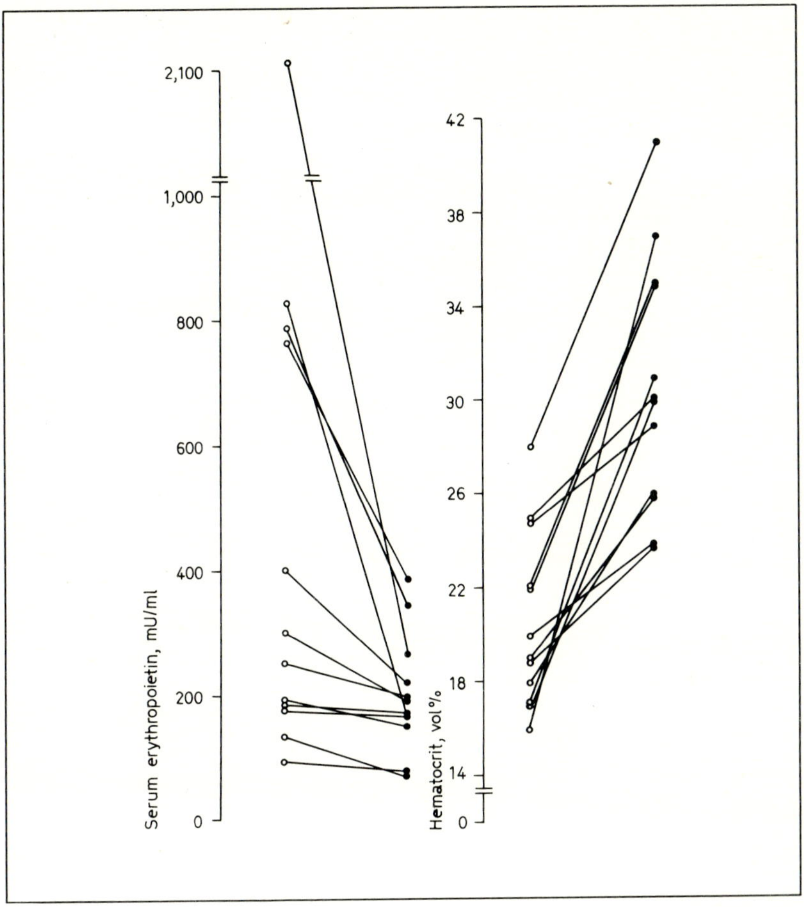

Fig. 4. Hematocrit and SEp concentration at the beginning (○) and after 3–27 months of regular hemodialysis treatment (●).

Again, two populations of patients become apparent when the data are further evaluated. In figure 5 pre- and posttreatment SEp concentrations are plotted against corresponding hematocrits. The patients with the highest levels of SEp at the start of treatment respond with the greatest decrease when hematocrit rises, while patients with lower SEp concentrations show only slight alterations in the SEp levels when anemia is partially corrected. Whereas in case of the latter type of patient Ep production appears to operate on a fixed maximum, the first group demonstrates an intact feedback mechanism between peripheral hemoglobin levels and Ep production. In view of this good compensatory ability

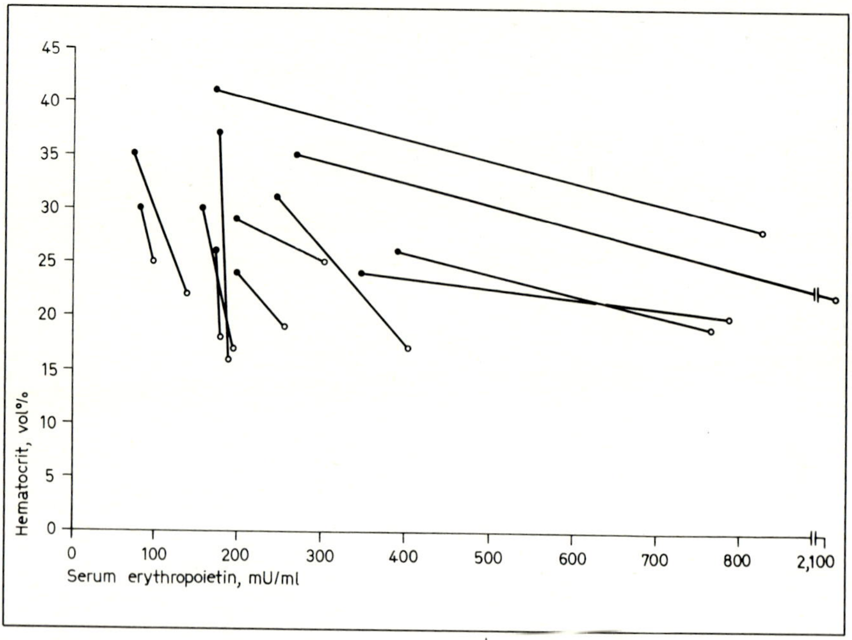

Fig. 5. Effect of regular hemodialysis treatment on hematocrit/erythropoietin relation. ○ = At beginning of treatment; ● = after 3–27 months of treatment.

of Ep-producing sites in a significant number of dialysis patients, it is not surprising when correction of the uremic state by regular hemodialysis treatment is often accompanied by a striking improvement of hemoglobin levels (25). It remains open whether this effect of hemodialysis is only a consequence of removal of still unknown toxins or whether other factors, such as improvement of diet and reduced hemolysis, are also involved. Reduction of blood loss can be excluded as a causative factor in this context, as total daily blood loss was proven to increase when patients in terminal renal failure are submitted to maintenance hemodialysis (26).

Summary

To evaluate the role of reduced Ep production in the pathogenesis of renal anemia, measurements of SEp concentration by means of a highly sensitive *in vitro* bioassay were performed in 100 nonnephrectomized, nontransfused regular hemodialysis patients. In 53 patients (group I) SEp levels were lower than 120 mU/ml. In the remaining 47 patients (group II) SEp concentrations ranged from 120 to 389 mU/ml. Group I showed a highly significant positive correlation between SEp concentration and hematocrit as did 13 anephric patients investigated for comparison. In contrast, group II displayed a highly significant negative correlation between SEp concentration and hematocrit. In 12 azotemic patients, where SEp concentration was measured on the first day of dialysis and 3–27 months following the onset of maintenance hemodialysis, again two populations of patients became apparent: With rising hematocrits, half of the patients responded with a significant drop of initially highly elevated SEp concentrations, whereas the other patients showed only insignificant changes in SEp levels when anemia was partially corrected by hemodialysis treatment. The results demonstrate the existence of two distinctive groups of similar size in patients with terminal renal failure: those with an absolute and those with a relative deficiency of Ep. In the case of the latter, lack of Ep is probably a secondary factor in the pathogenesis of anemia, whereas uremic toxicity, blood loss and malnutrition may play a primary role.

References

1 *Erslev, A.J.:* Anemia of chronic renal disease. Archs intern. Med. *126:* 774–780 (1970).

2 *Fried, W.:* Erythropoietin and the kidney. Nephron *15:* 327–349 (1975).

3 *Eschbach, J.W.; Funk, P.; Adamson, J.; Kuhn, J.; Scribner, B.H., and Finch, C.A.:* Erythropoiesis in patients with renal failure undergoing chronic dialysis. New Engl. J. Med. *276:* 653–658 (1967).

4 *Zucker, S.; Lysik, R.M., and Mohammad, G.:* Erythropoiesis in chronic renal disease. J. Lab. clin. Med. *88:* 528–535 (1976).

5 *Ypersele de Strihou, C. van and Strgier, A.:* Effect of bilateral nephrectomy on transfusion requirements of patients undergoing chronic dialysis. Lancet *ii:* 705 (1969).

6 *Fried, W.; Jonasson, O.; Lang, G., and Schwartz, F.:* The hematologic effect of androgen in uremic patients. Ann. intern. Med. *79:* 823–827 (1973).

7 *Naetes, J.P. and Heuse, A.F.:* Measurement of erythropoietic stimulating factor in anemic patients with or without renal disease. J. Lab. clin. Med. *60:* 367–374 (1972).

8 *Erslev, A.J.:* The effect of uremic toxins on the production and metabolism of erythropoietin. Kidney int., suppl., pp. 129–133 (1975).

9 *Wallner, S.F.; Ward, H.P.; Vautrin, R.; Alfrey, A.C., and Mishell, J.:* The anemia of chronic renal failure. *In vitro* response of bone marrow to erythropoietin. Proc. Soc. exp. Biol. Med. *149:* 939–944 (1975).

10 *Gurney, C.W.; Goldwasser, E., and Pan, C.:* Studies on erythropoiesis. IV. Erythropoietin in human plasma. J. Lab. clin. Med. *50:* 534 (1961).

11 *Gallagher, N.I.; McCarthy, J.M.; Hart, K.T., and Lange, R.D.:* Evaluation of plasma erythropoietic stimulating factors in anemic uremic patients. Blood *14:* 662–667 (1959).

12 *De Gowin, R.L.; Lavender, A.R.; Forland, M.; Charleston, D., and Gottschalk, A.:* Erythropoiesis and erythropoietin in patients with chronic renal failure treated with hemodialysis and testosterone. Ann. intern. Med. *72:* 913–918 (1970).

13 *Lertora, J.J.L.; Dargon, P.A.; Rege, A.B., and Fisher, J.W.:* Studies on a radioimmunoassay for human erythropoietin. J. Lab. clin. Med. *86:* 140–151 (1975).

14 *Schulz, E.; Mödder, B. und Rath, K.:* Verhalten von Plasma-Erythropoietin (ESF) und Hämatokrit unter dem Einfluss der Dauerdialyse-Behandlung. Klin. Wschr. *55:* 65–69 (1977).

15 *Blumberg, A.:* Die renale Anämie (Huber, Bern 1972).

16 *Davies, S.; Glynne-Jones, E.; Bisson, M., and Bisson, P.:* Plasma erythropoietin assay in patients with chronic renal failure. J. clin. Path. *28:* 875–878 (1975).
17 *Essers, U.:* Erythropoietinaktivität und Bedarf an Bluttransfusionen bei Langzeit-dialyse-Behandlung. Blut *24:* 346–349 (1972).
18 *Eschbach, J.W.; Adamson, J.W., and Cook, J.D.:* Disorders of red blood cell production in uremia. Archs intern. Med. *126:* 812–815 (1970).
19 *Dunn, C.D.R.; Jarvis, J.H., and Greenman, J.M.:* A quantitative bioassay for erythropoietin using mouse fetal liver cells. Expl Hematol. *3:* 65–78 (1975).
20 *Wardle, D.F.H.; Baker, I.; Malpas, J.S., and Wrigley, P.F.M.:* Bioassay for erythropoietin using fetal mouse liver cells. Br. J. Haemat. *24:* 49–56 (1973).
21 *Annable, L.; Cotes, P.M., and Mussett, M.V.:* Second international reference preparation of erythropoietin, human, urinary for bioassay. Bull. Wld Hlth Org. *47:* 99–112 (1972).
22 *Colquhoun, D.:* Lectures on biostatistics. An introduction to statistics with application in biology and medicine, pp. 327–332 (Clarendon Press, Oxford 1971).
23 *Blumberg, A. und Keller, H.:* Die Erythropoiese bei anephrischen Patienten. Schweiz. med. Wschr. *101:* 1887–1893 (1971).
24 *Murphy, G.P.; Mirand, E.A., and Kenny, G.M.:* Renal and extrarenal erythropoietin in man. N.Y. St. J. Med. *1969:* 2007–2011.
25 *Koch, K.M.; Patyna, W.D.; Shaldon, S., and Werner, E.:* Anemia of the regular hemodialysis patients and its treatment. Nephron *12:* 405–419 (1974).
26 *Koch, K.M.; Werner, E.; Bechstein, P. und Fassbinder, W.:* Die Bedeutung occulter Blutverluste für die Eisenbilanz des chronischen Hämodialysepatienten; in *von Dittrich, Skrabal und Stühlinger* Aktuelle Probleme der Dialyseverfahren und der Niereninsuffizienz. V. Symp. Innsbruck, pp. 397–405 (Bindernagel, Friedberg 1975).

Dr. *K.M. Koch,* Department of Nephrology, Center of Internal Medicine, University of Frankfurt, Theodor-Stern-Kai 7, *D-6000 Frankfurt (Main) 70* (FRG)

Contr. Nephrol., vol. 13, pp. 69–80 (Karger, Basel 1978)

Role of Erythropoietin in the Anemia of Renal Insufficiency in Man and in an Experimental Uremic Rabbit Model

E. Schulz, B. Mödder and J.W. Fisher

Medizinische Universitätspoliklinik und Medizinische Klinik
(Direktor: Prof. *W. Kaufmann*), Köln-Merheim, and Department of Pharmacology,
Tulane University Medical School, New Orleans, La.

Anemia of chronic renal failure is a complex disease entity. Several factors which are postulated to play a role in the mechanism of the anemia of renal insufficiency are listed in table I.

Most of the earlier studies performed have demonstrated that anemia of uremia is due to some extent to hemolysis and to a hypoproliferative erythropoietic state of the bone marrow. The present studies are concerned with the role of decreased production of erythropoietin (ESF) and of a possible decreased effect of ESF on the bone marrow itself.

The possible significance of ESF in the mechanism of the anemia of uremia is closely linked to the site of ESF production. Reduced kidney oxygen tension may trigger the production and release of the renal erythropoietic factor (REF) activating the plasma factor proerythropoietin to generate ESF. ESF then acts on the bone marrow to stimulate erythropoiesis. It is conceivable that in renal failure this control mechanism is altered at a very sensitive site, namely the

Table I. Pathogenetic factors of anemia in chronic renal disease

A Hemolytic factor
- 1 Extracellular factor (uremic toxins, acidosis) intracellular factor
- 2 Mechanical hemolysis

B Decreased erythropoiesis
- 1 Decreased production of erythropoietin and/or inhibitors of erythropoietin
- 2 Uremic toxic effect on erythropoietic cells
- 3 Decreased efficiency of erythropoietin on erythropoietic cells

C Anemia of chronic infection

D Blood loss

E Iron deficiency

F Folic acid deficiency

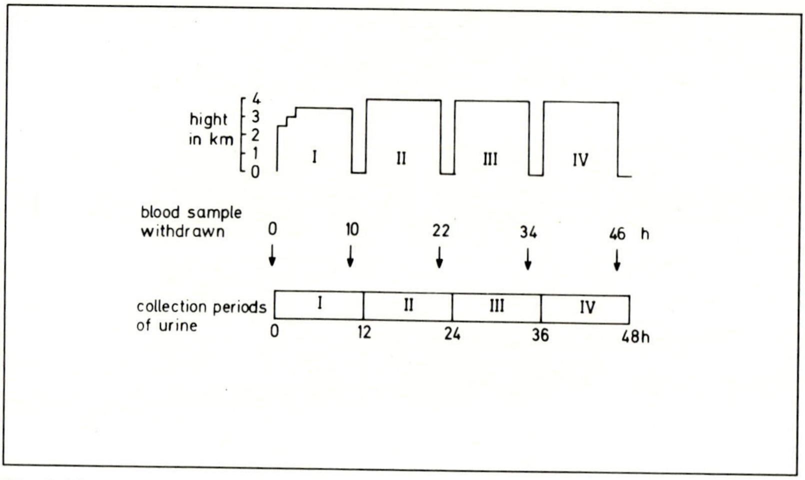

Fig. 1. Time course of hypobaric chamber experiments.

kidney itself, causing a decreased production of REF. Therefore, decreased production of ESF in chronic renal failure may be a primary cause of the decreased erythropoiesis in this condition. We performed studies on ESF in patients with mild and severe renal insufficiency combined with anemia of various degrees of severity.

ESF in Incipient Renal Failure

It seems of interest to learn whether a latent ESF deficiency is present in incipient anemia of uremia in patients during hypoxic stimulation. ESF production was stimulated via hypoxia in 8 nonanemic patients suffering from chronic renal failure having a creatinine clearance of 38 ± 3 ml/min (mean ± SEM) as well as in 8 normal human subjects. Hypoxia was produced in a hypobaric chamber at a reduced pressure corresponding to an altitude of 4,000 m above sea level for a 4 × 10 h period (fig. 1). Serum and urine ESF titers were determined using the exhypoxic polycythemic mouse assay (1, 2). During hypoxic stimulation serum ESF titers in patients with chronic renal failure were significantly higher after 10 h of hypoxia than in normal human subjects (fig. 2). Urine excretion showed a similar pattern, the difference, however, was not significant (fig. 3). It is concluded that a latent ESF deficiency cannot account for the incipient anemia in chronic renal failure. The increased serum ESF titers in patients with chronic renal failure during hypoxic stimulation may be due to the lower hematocrit values (mean = 42%) compared to normal individuals (mean = 46%).

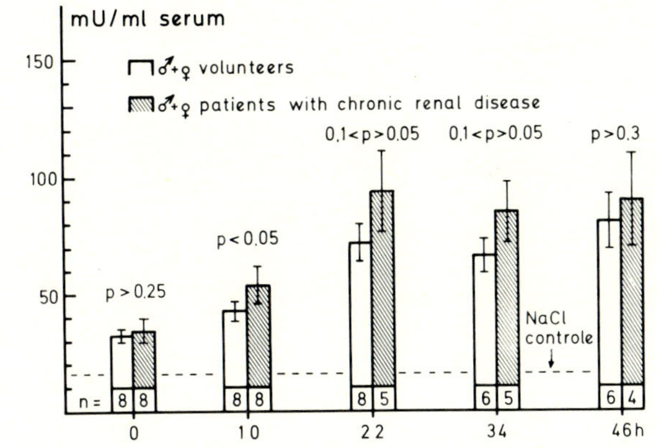

Fig. 2. Serum ESF level (mean ± SEM) before and after hypoxic stimulation. n = Number of probands; p = estimation of significance by Student's test.

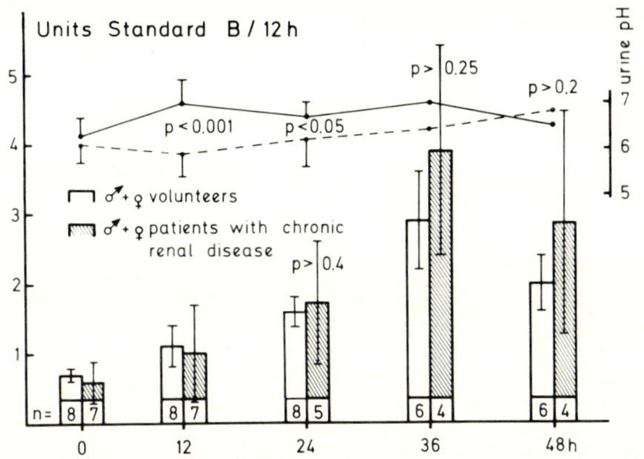

Fig. 3. ESF excretion (mean ± SEM) before and during hypoxic stimulation in 12-hour urine samples, urine pH in volunteers (——) and in patients (– – –).

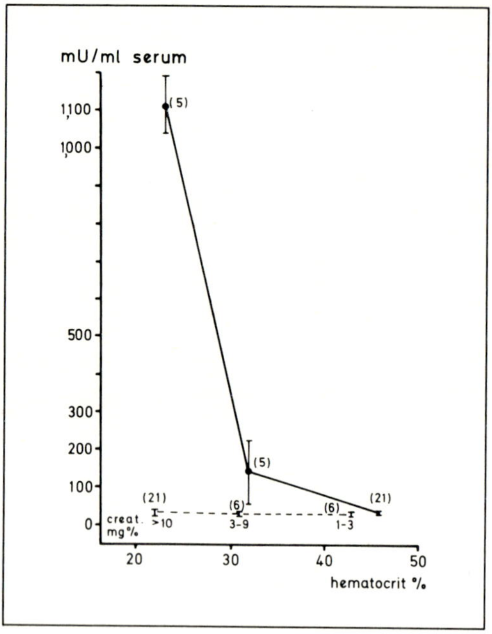

Fig. 4. Serum ESF titer and hematocrit in patients with normal renal function (——) and in chronic renal failures (– – –) (mean ± SEM). Number of patients in parentheses.

ESF in Severe Anemia due to Advanced Chronic Renal Failure

In patients with normal renal function (e.g. aplastic anemia) increased titers of ESF have been demonstrated in which the increase is proportional to the decreased hematocrits. In our studies serum ESF titers in normal subjects (mean hematocrit = 46%) were 34 ± 2 mU/ml. Patients with normal renal function had serum ESF titers of 142 ± 76 mU/ml when mildly anemic and 1,115 ± 74 mU/ml when severely anemic. On the other hand, there was no increase in serum ESF in patients with anemia of uremia (fig. 4). Therefore, a relative ESF deficiency may be an important factor in the anemia of chronic renal failure, since serum ESF titers are lower in advanced renal failure than in individuals with comparable degrees of anemia induced by other illnesses.

It has been demonstrated by *Blumberg et al.* (3) and by *Blumberg* (7) that patients without any excretory renal function undergoing chronic intermittent hemodialysis still had increased serum ESF titers when exposed to hypoxic stimulation. These results indicate persisting regulation of ESF production.

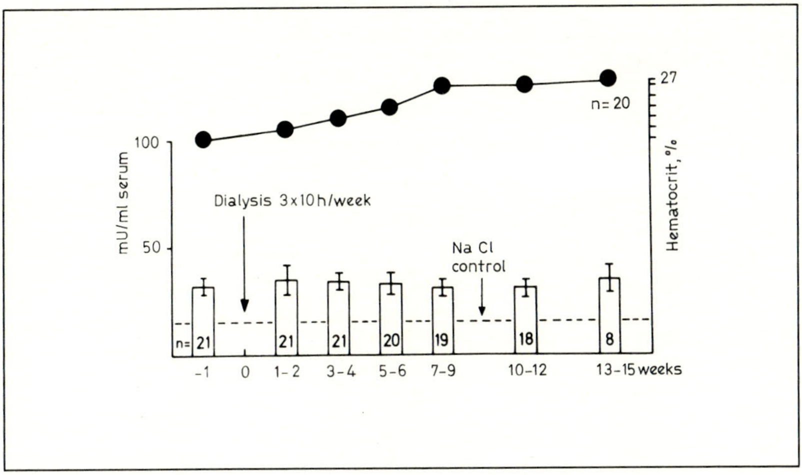

Fig. 5. Plasma ESF and hematocrit under the influence of chronic hemodialysis treatment (mean ± SEM).

Plasma ESF Titers and Hematocrits during Chronic Hemodialysis

Berry et al. (4) have demonstrated in 1964 that erythropoiesis improves following chronic dialysis treatment. Figure 5 demonstrates increasing hematocrits in our home dialysis patients over a period of 15 weeks after initiation of dialysis. Hematocrits increased from 20 to 34% within the first 2 years of dialysis. The question arises whether dialysis increases ESF production. Figure 5 also demonstrates ESF titers in 21 home dialysis patients. While hematocrits were increased significantly, plasma ESF titers did not change over the period studied. These results suggest that the improved erythropoiesis followed by the increase in hematocrits does not depend upon enhanced ESF production (5).

ESF and Erythropoiesis after Bilateral Nephrectomy (Extrarenal ESF in Man)

It is of interest to note that after bilateral nephrectomy which reflects extrarenal ESF production, a complete disappearance of erythroblasts has been demonstrated by *Naets et al.* (6) after bilateral nephrectomy within 3 days in dogs. This phenomenon is less marked in man. Normoblasts were decreased from 13.9 to 6% followed by an increase to 10% within 30 days (6). Unlike the dog complete erythroblastophthisis does not occur in man. Hematocrits were de-

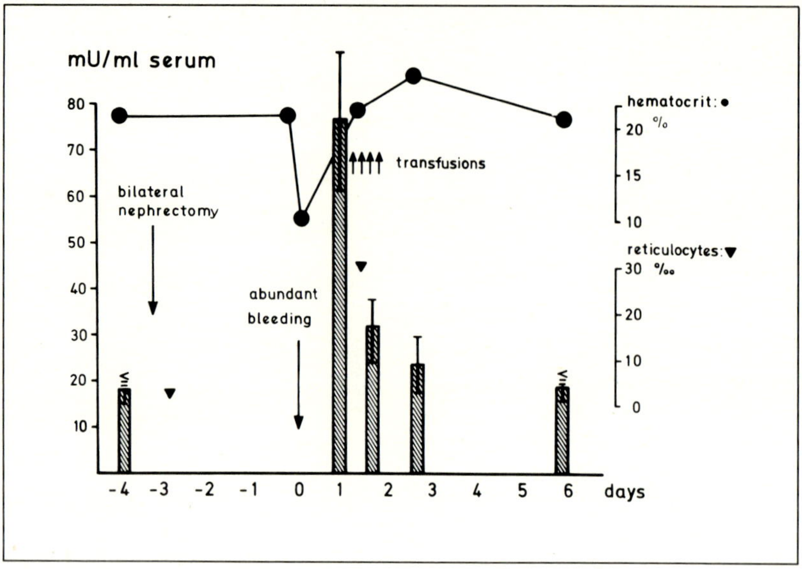

Fig. 6. Plasma ESF titer, reticulocytes, and hematocrit after bilateral nephrectomy and abundant bleeding (mean ± SEM).

creased within 2 months after bilateral nephrectomy, followed by a gradual increase (7). Obviously, human subjects are capable of extrarenal ESF production as are primates (8). Extrarenal ESF production is also triggered by hypoxia as indicated by our observation of a bilateral nephrectomized patient suffering from an acute blood loss. Before the bleeding episode the hematocrit was 21% and the serum ESF titers were not detectable (below 18 mU/ml). When the hematocrit was reduced to 10%, serum ESF titer increased 4-fold to 77 mU/ml followed by a gradual decrease after blood transfusion (fig. 6).

ESF after Renal Transplantation

Hemoglobin concentration is generally increased after successful renal transplantation (9) and is usually higher than in renal insufficiency. *Gral et al.* (10) were able to demonstrate detectable ESF titers before renal transplantation only when hematocrits were below 20%. However, serum ESF was detectable within 24 h after renal transplantation and was followed by a gradual decrease over 2 months while the hematocrit was increased above 35%. Unlike normal human subjects there was no correlation between serum ESF titers and hematocrits in renal transplant recipients as described by *Mirand and Murphy* (11). In order to

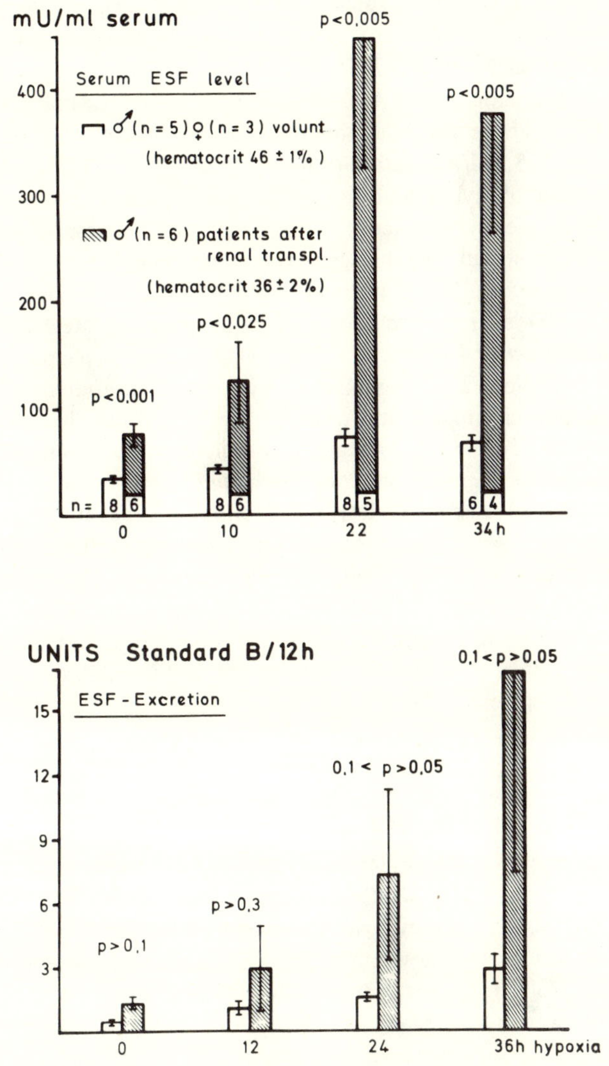

Fig. 7. Serum ESF titer and ESF excretion (mean ± SEM) in renal transplanted male and female subjects before and under hypoxic stimulation.

study ESF production in transplanted kidneys, serum ESF titers and urine excretion were determined with and without hypoxic stimulation using a hypobaric chamber in 8 normal subjects and 6 renal transplant recipients (fig. 7). Serum creatinine in transplant recipients was between 0.9 and 1.6 mg% and creatinine clearance between 80 and 143 ml/min. The mean venous hematocrit was 36% in the patients and 46% in the normal subjects. Under normal ambient pressure (57 m above sea level) the mildly anemic patients had serum ESF titers twice as high (75 mU/ml) as that of normal individuals (34 mU/ml). 22 and 34 h after initiation of hypoxia, respectively, serum ESF titers in normal human subjects were merely doubled (34 to 72–67 mU/ml, respectively). The additional hypoxic stimulation in renal transplant recipients, however, was more than additive: serum ESF titers were 5-fold increased 22 h after initiation of the experiment. ESF excretion was correspondingly elevated. Mean ESF excretion in normal subjects was 2.9 and 13.4 U in renal transplant recipients between the 24th and 36th hour after beginning of hypoxia. The difference in the hematocrits in the normal subjects and transplant recipients may be the reason for the difference seen. The results indicate, however, that the denervated transplanted kidney is capable of producing considerable amounts of ESF due to the anemia of the transplant recipient. ESF production can be stimulated further by hypoxic stimulation. Therefore, the mild anemia in the 6 renal transplant recipients seems not to be due to ESF deficiency. On the other hand, uremic toxins cannot be responsible for the anemia, since creatinine clearances were between 80 and 140 ml/min. Therefore, the persisting anemia may be due to immunosuppressive therapy with azathioprine and its inhibitory effects on bone marrow erythropoiesis (12).

Decreased Effectiveness of ESF on Bone Marrow

An increase in reticulocytes by 80–100% has been demonstrated in normal human subjects 36 h after administration of 500 U ESF. There was no increase in reticulocytes seen, however, in 4 uremic patients under the same conditions (13). These results suggest suppression of the bone marrow in uremia. We have studied this problem further in an experimental uremic animal model.

Rabbits were made chronically uremic by removing five sixths of the kidney mass. In the first operation two thirds of one kidney was removed followed by a total nephrectomy of the contralateral kidney 14 days later. The second operation was followed by a sharp increase of blood urea nitrogen (BUN) and creatinine. The serum was assayed for ESF in the exhypoxic polycythemic mouse. None of the chronic uremic or sham-operated rabbits had detectable serum titers of ESF. Bone marrow cell suspensions were cultured *in vitro* to determine the relative number of CFU-E (colony forming unit-erythroid) in the

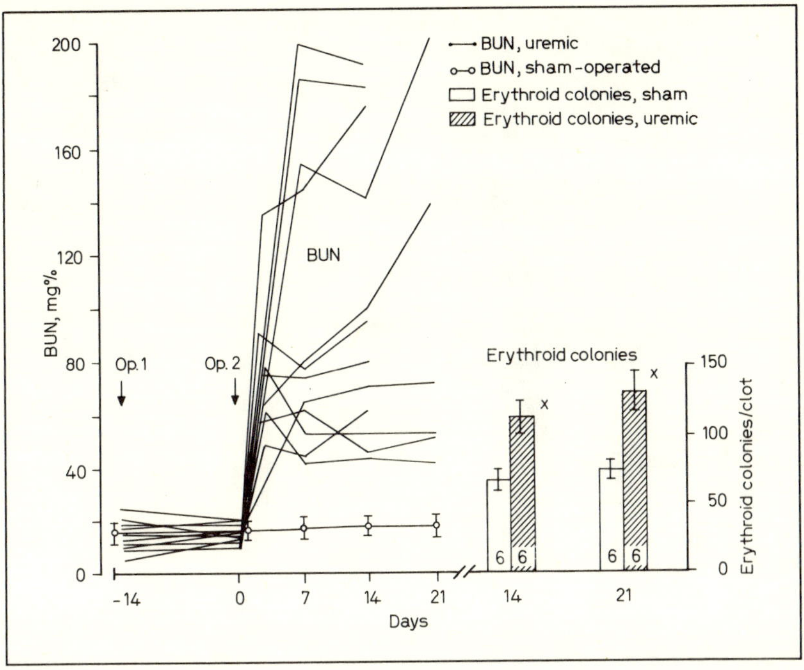

Fig. 8. Erythroid colony formation in chronic uremic rabbit bone marrow cultures (group 1). Each bar is the mean ± SEM and the number at the bottom of each bar indicates the number of experiments. X = Significantly different from control ($p < 0.01$).

bone marrow using a modification of the plasma clot erythroid colony forming cell technique of *McLeod et al.* (14). According to the course of the uremia the animals were divided into two groups. Group 1 consisted of rabbits having permanently elevated or continuously increasing BUN and creatinine. Group 2 consisted of rabbits with decreasing BUN and creatinine after an initial increase. Each erythroid colony growing in the bone marrow culture is postulated to reflect one CFU-E (a cell which responds to ESF) within the bone marrow cells cultured. Therefore, the number of erythroid colonies scored after incubation of the cultures reflects the number of CFU-E in the uremic marrow and the marrow of sham-operated controls.

Figure 8 shows the results of the five-sixth nephrectomized rabbits in group 1. After 14–21 days of uremia rabbits in this group had significantly higher numbers of CFU-E in their bone marrows than controls. Group 2 rabbits (fig. 9), having a decreasing BUN after an initial increase, had significantly less erythroid colonies in their bone marrow cultures. Figure 10 shows the hematocrits of the uremic rabbits in groups 1 and 2 which were significantly lower than controls,

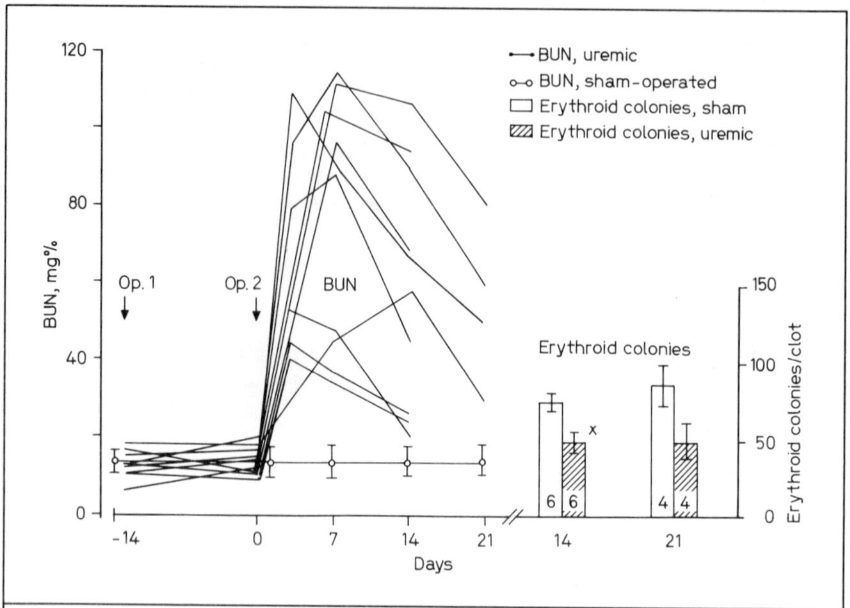

9

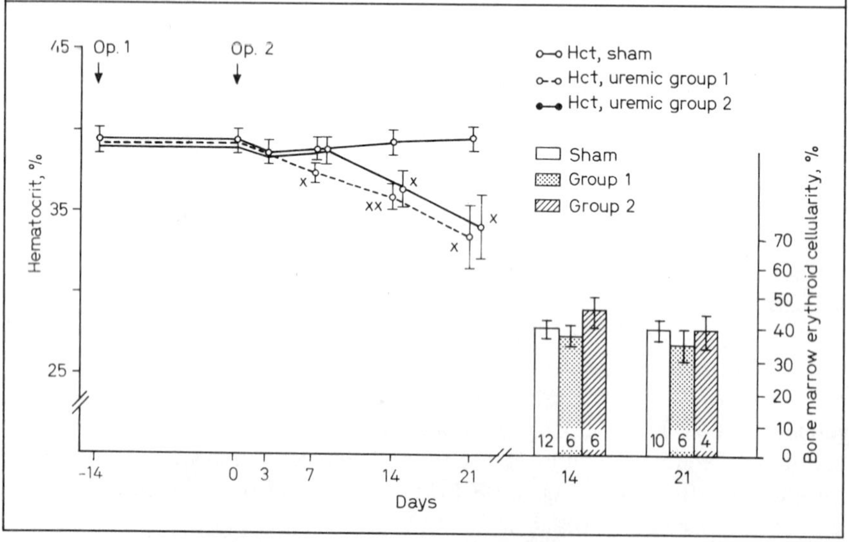

10

Fig. 9. Erythroid colony formation in chronic uremic rabbit bone marrow cultures (group 2). Each bar is the mean ± SEM and the number at the bottom of each bar indicates the number of experiments. × = Significantly different from control ($p < 0.02$).

Fig. 10. Hematocrit (Hct) and bone marrow erythroid cellularity in chronic anemic uremic rabbits following five-sixth nephrectomy. Each bar is the mean ± SEM and the number at the bottom of each bar indicates the number of experiments. × = Significantly different ($p < 0.05$) from sham; ×× = $p < 0.01$.

indicating a developing anemia of renal insufficiency. As one might expect, the more severely uremic rabbits of group 1 had somewhat lower hematocrits than group 2. Bone marrow erythroid cellularity in groups 1 and 2 was not significantly different from that of controls.

The following conclusions can be made. Uremia increases the number of CFU-E in the bone marrow. There may be a blockade in the further differentiation into morphologically recognizable erythroid precursor cells, thereby contributing to the hypoproliferative state of the bone marrow. Due to some regeneration of the kidney remnant in group 2 rabbits, improving kidney function may have been responsible for the decreasing BUN and creatinine within the last 7–14 days studied. It is conceivable that the blockade of the early erythroid cells in these rabbits was released, followed by a rapid outflow of the CFU-E compartment due to the increased demand for erythrocytes. Therefore, the number of CFU-E of group 2 rabbits was lower than controls.

The nature of the factor(s) inhibiting further differentiation of the CFU-E is not known. A relative lack of ESF may play a role as well as factors inhibiting the action of ESF on the bone marrow.

Summary

ESF deficiency is probably not a major contributing factor in the early stages of the anemia of renal insufficiency. Serum ESF titers are lower in advanced renal failure when compared to that of nonuremic anemic subjects suffering from equivalent anemia. With increasing renal insufficiency a relative ESF deficiency gains increasing importance as a pathogenic factor in reduced erythropoiesis.

Kidneys without excretory function may still be erythropoietically effective, since a further increase in the anemia occurs after bilateral nephrectomy. However, a basal erythropoiesis is still maintained by extrarenal ESF production, which is also enhanced by hypoxia. ESF deficiency is compensated after successful renal transplantation.

A decreased response of the bone marrow to ESF may be another factor contributing to the hypoproliferative state of erythropoiesis in uremia. As demonstrated in a chronic uremic rabbit model there may be a blockade of further differentiation of the erythroid precursors. The relationship of this blockade in differentiation to the inhibitor of heme synthesis is not clear.

References

1 *Schulz, E. and Cissée, H.:* Comparative experiments on erythropoietin concentrations in ultrafiltered urine (Diaflo-process) and urine dialysis with higher osmolar substances (Carbowax). J. clin. Chem. clin. Biochem. *14:* 537–542 (1976).

2 *Schulz, E. and Cissée, H.:* Excretion of erythropoietin by humans during the production of alkaline or acidic urine. J. clin. Chem. clin. Biochem. *14:* 557–559 (1976).

3 *Blumberg, A.; Keller, H., and Marti, H.R.:* Effect of altitude on erythropoiesis and oxygen affinity in anaemic patients on maintenance dialysis. Eur. J. clin. Invest. *3:* 93–97 (1973).

4 *Berry, E.R.; Rambach, W.A.; Alt, H.L., and Greco, F. del:* Effect of peritoneal dialysis on erythrokinetics and ferrokinetics of azotemic anemia. Trans. Am. Soc. Artif. internal Organs *10:* 414 (1964).

5 *Schulz, E.; Mödder, B. und Rath, K.:* Verhalten von Plasma-Erythropoetin (ESF) und Hämatokrit unter dem Einfluss der Dauerdialyse-Behandlung. Klin. Wschr. *55:* 65–69 (1977).

6 *Naets, J.; Wittek, M.; Toussaint, C., and Gerstruyden, J. van:* Erythropoiesis in renal insufficiency and anephric man. Ann. N.Y. Acad. Sci. *149:* 143 (1968).

7 *Blumberg, A.:* Die renale Anämie (Huber, Bern 1972).

8 *Mirand, E.A.; Steeves, R.A.; Groenewald, J.H.; Zyl, J.J.W. van, and Murphy, G.P.:* Extrarenal erythropoietin production in the baboon. Proc. Soc. exp. Biol. Med. *130:* 685 (1969).

9 *Pichlmayr, R. und Grotelüschen, B.:* Entwicklung und heutiger Stand der Nierentransplantation. Medsche Welt *24:* 26 (1973).

10 *Gral, J.; Schroth, P., and Palma, J.R. de:* Kidney graft failure, erythropoietin (EP), and immunosuppression. Abstr. VIIth Congr. Eur. Dialysis Transplant. Ass., Barcelona 1970.

11 *Mirand, E.A. and Murphy, G.P.:* Erythropoietin alterations in patients with uremia, renal allografts, or without kidneys. J. Am. med. Ass. *209:* 392 (1969).

12 *Jepson, J.H. and Manase, B.:* Enhanced response of erythropoietin by low dose prednisolone hemisuccinate and its abolition by azathioprine. Biomedicine *20:* 179–185 (1974).

13 *Essers, U.; Müller, W. und Brunner, E.:* Zur Wirkung von Erythropoietin bei Gesunden und bei Patienten mit chronischer Urämie. Klin. Wschr. *51:* 1005 (1973).

14 *McLeod, D.L.; Shreeve, M.M., and Axelrad, A.A.:* Plasma culture system for production of erythrocytic colonies *in vitro*. Blood *44:* 517 (1974).

Dozent Dr. *E. Schulz,* Innere Abteilung, Vinzenz-Pallotti-Hospital,
D–5060 Bergisch Gladbach 1 (FRG)

Vitamin D

Contr. Nephrol., vol. 13, pp. 81–95 (Karger, Basel 1978)

Regulation of the Vitamin D Endocrine System Located in the Kidney[1]

Hector F. DeLuca

Department of Biochemistry, College of Agricultural and Life Sciences, University of Wisconsin-Madison, Madison, Wisc.

During the past decade has come the realization that vitamin D is in effect a prohormone which is converted by sequential activation in the liver and in the kidney to an active hormone which then functions to regulate calcium and phosphorus metabolism (1–4). It is well known that in the absence of vitamin D newly laid down organic matrix of bone fails to mineralize, giving rise to the disease osteomalacia of the adult and rickets in children (5). Thus it has been assumed that vitamin D must function directly on the mineralization system located in the chondroblasts and osteoblasts. So far evidence to support this idea is lacking although it still remains an intellectually viable hypothesis. However, it seems abundantly clear that the primary reason for defective mineralization of newly laid down organic matrix of bone in vitamin D deficiency is an insufficient supply of calcium and phosphorus to the mineralization centers from the extracellular fluid compartment (6). Thus in vitamin D deficiency, plasma calcium times phosphorus concentrations are undersaturated (7) with regard to the hydroxyapatite mineral component of bone whereas normally blood is supersaturated with these two ion components (8). Therefore a large if not the only measure of vitamin D function in the mineralization of bone is the elevation of extracellular fluid compartment calcium and phosphorus to levels that will support mineralization (6).

In addition to the failure of mineralization, it is well known that a deficiency of vitamin D will bring about the condition of hypocalcemic tetany even in the presence of normal parathyroid glands (9). It is, therefore, evident that vitamin D plays an important role in the prevention of hypocalcemic tetany. Obviously the prevention of this disease is by the elevation of plasma calcium concentration. Therefore, the basic function of vitamin D is to elevate plasma

[1] Some of the original investigations reported in this paper were supported by US Public Health Service Grant AM-14881 and the Harry Steenbock Research Fund.

calcium and phosphorus concentrations. This is brought about by the stimulation of calcium and phosphorus transport across intestinal epithelium (10–13), improved renal reabsorption of calcium (14, 15) and increased mobilization of calcium from the bone fluid compartment. All of these systems contribute calcium and phosphorus to the extracellular fluid compartment preventing hypocalcemic tetany on one hand and causing the mineralization of bone on the other. Finally, it is important to realize that the intestine is able to adjust to the calcium need of the organism by elevating its efficiency of intestinal calcium absorption (16). The elevation of intestinal calcium absorption in response to the need for calcium requires vitamin D (16). Thus an important aspect of vitamin D function is to permit the intestine to adjust its calcium absorption according to the organismal needs (17). If this failure to adjust intestinal calcium absorption does not take place, the physiology of man and higher animals is so constructed that the skeleton will be sacrificed to support serum calcium concentration contributing, therefore, to the erosion of bone or osteoporosis. It has been demonstrated that the mechanism whereby the intestine adjusts its absorption of calcium according to needs is via the parathyroid hormone-vitamin D metabolism system as will be discussed below (17, 18).

In all of these above-described functions of vitamin D, it is now well established that the following metabolic alterations in the vitamin D molecule must occur prior to carrying out the functions just described. As demonstrated in figure 1, 7-dehydrocholesterol of the epidermis is acted upon by 300-nm ultraviolet light to produce vitamin D_3 (19). Recent work in our laboratory has demonstrated that in fact the product of ultraviolet irradiation of skin is vitamin D_3 ultimately (20). Inasmuch as insufficient amounts of ultraviolet light reach the epidermis of most individuals, vitamin D has become a dietary essential

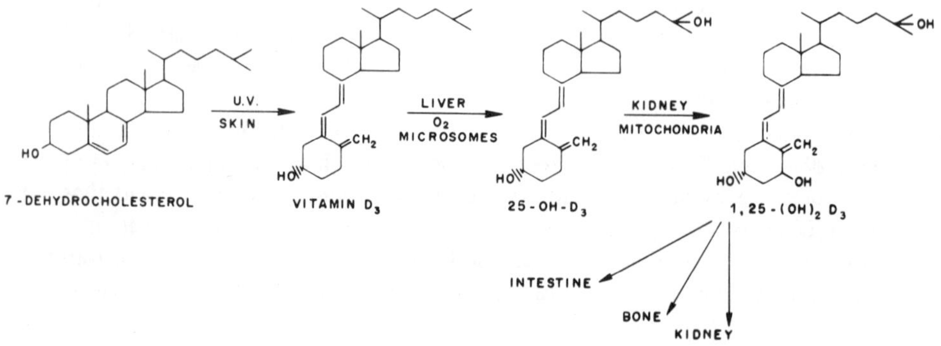

Fig. 1. Metabolic alterations of vitamin D which are required for physiologic functions.

(21). Thus vitamin D can either be manufactured in the skin or it can be absorbed from the intestinal tract. Upon its appearance in the plasma, vitamin D is rapidly taken up by the liver with as much as 80% of a dose accumulating in the liver within 60 min (22). In the liver vitamin D undergoes its first obligatory metabolic alteration in which a hydroxyl is inserted on carbon 25 to form 25-hydroxyvitamin D_3 (25-OH-D_3) (23–25). This reaction occurs in the endoplasmic reticulum (26) and is regulated by the product itself (27) although this regulation can be overcome in an as yet undefined manner by increasing the vitamin D reaching the liver (25, 28). 25-OH-D_3, although 2–5 times more active than the parent vitamin D in the physiologic functions described above (29), does not act directly in any target tissue at physiologic concentrations. It is the major circulating form of vitamin D bound to a 52,000 molecular weight α_2-globulin which is considered at the present time a specific transport protein for the vitamin D compounds (30–32). 25-OH-D_3 is taken up by the kidney where it undergoes its second obligatory metabolic alteration. In the mitochondria (33) of presumably the proximal tubule epithelial cells 25-OH-D_3 is converted to the active hormone 1,25-dihydroxyvitamin D_3 (1,25-[OH]$_2D_3$) (34). This reaction occurs exclusively in the kidney (35, 36) and exclusively in the mitochondria (33). The reaction is catalyzed by a three-component system involving a flavoprotein, an iron sulfure protein and a specific cytochrom P-450 enzyme which actually carries out the hydroxylation on carbon 1 (37–39). The 1,25-$(OH)_2D_3$ is then delivered to the intestine, bone and distal tubule where it carries out the functions described. Nephrectomy prevents the response of intestine and bone to physiologic concentrations of 25-OH-D_3 whereas it does not prevent the response of these target organs to 1,25-$(OH)_2D_3$ (40–42). Furthermore, experiments with radioactive 1,25-$(OH)_2D_3$ reveal that it is likely the active form in stimulating intestinal calcium transport and in stimulating the mobilization of calcium from bone (43–45). It is possible that 1,25-$(OH)_2D_3$ may be furhter metabolically altered before it can carry out the function in the phosphate transport mechanisms; so far that has not yet been resolved (1, 46). Based on the above experiments, 1,25-$(OH)_2D_3$ being manufactured exclusively in the kidney and having its function in the targets of intestine and bone can be considered a true hormone with the kidney as the endocrine organ (47). More specifically, the vitamin D system in the kidney is a true endocrine organ since the biogenesis of 1,25-$(OH)_2D_3$, a calcium-regulating hormone, is regulated by the need for calcium (48–50), the need for phosphorus (51) and the parathyroid hormone (52). Prior to considering these points, it is important to realize that the kidney manufactures another important metabolite of 25-OH-D_3, namely 24, 25-dihydroxyvitamin D_3 (24, 25-[OH]$_2D_3$) (53, 54). The function of the 24-hydroxylase which acts on any vitamin D molecule possessing a 25-hydroxyl group remains unknown. However, it is absent in the vitamin D-deficient state and can be induced by 1,25-$(OH)_2D_3$

(55, 56). In general its activity appears to be regulated in the reverse direction from the 1-hydroxylase (57). It has recently been demonstrated in our laboratory that in addition to the kidney, the 24-hydroxylase is also found in the intestine (58). Thus nephrectomy, although diminishing the amount of 24,25-$(OH)_2D_3$ appearing in the plasma, does not eliminate production of this metabolite or the metabolite which results from 24-hydroxylation of 1,25-$(OH)_2D_3$ (59). In general the view presently is held that the 24-hydroxylated vitamins are much less active than their non-24-hydroxylated counterparts which suggests that they may well be the first event involved in the inactivation of the active hormone 1,25-$(OH)_2D_3$ or the inactivation of 25-OH-D_3 (60, 61). It should also be mentioned that the exact stereochemistry of the 24 position has been shown to be *R* (59, 60). Furthermore, all of the above metabolites have been chemically synthesized and are available for experimental work and in some cases clinical investigation (61).

The biogenesis of 1,25-$(OH)_2D_3$ is regulated by the need for calcium as illustrated by a plot of serum calcium concentration versus the conversion *in vivo* of radioactive 25-OH-D_3 to either 24, 25-$(OH)_2D_3$ or 1,25-$(OH)_2D_3$ (fig. 2) (48, 49). Under conditions of normal calcemia some 1,25-$(OH)_2D_3$ is made but as the animal becomes hypoalcemic there is a stimulation of biogenesis of the 1,25-$(OH)_2D_3$ and a suppression of the 24-hydroxylation of 25-OH-D_3 (50, 62). On the other hand, hypercalcemia brings about a suppression of 1-hydroxylation

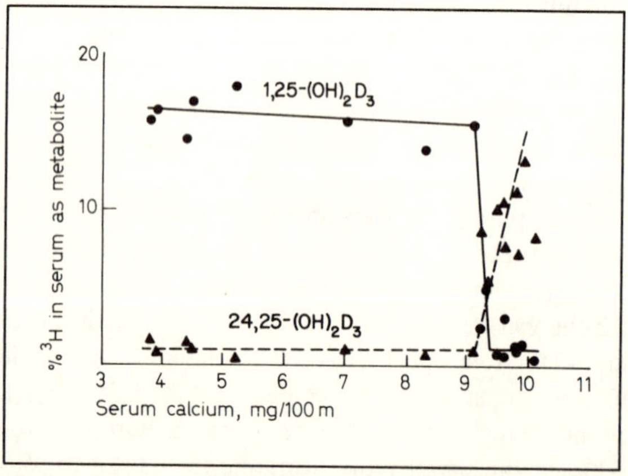

Fig. 2. The relationship between serum calcium concentration and the accumulation of 1,25-$(OH)_2D_3$ or 24,25-$(OH)_2D_3$ from its precursor 25-OH-D_3. Note that at normal calcemia both 24,25-$(OH)_2D_3$ and 1,25-$(OH)_2D_3$ are accumulated. Under conditions of hypocalcemia increased levels of 1,25-$(OH)_2D_3$ are obvious together with a marked suppression of 24,25-$(OH)_2D_3$ levels. Under conditions of hypercalcemia little or no 1,25-$(OH)_2D_3$ accumulates while large amounts of 24,25-$(OH)_2D_3$ appear.

and a stimulation of 24-hydroxylation. Thus the need for calcium in some manner stimulates biogenesis of the 1,25-$(OH)_2D_3$, the calcium mobilizing hormone. A series of experiments in our laboratory (52) and by *Fraser and Kodicek* (63) has demonstrated that the mechanism involved in stimulation of 1,25-$(OH)_2D_3$ biosynthesis in response to hypocalcemia is mediated by the parathyroid gland (fig. 3). It is well known that the parathyroid glands monitor

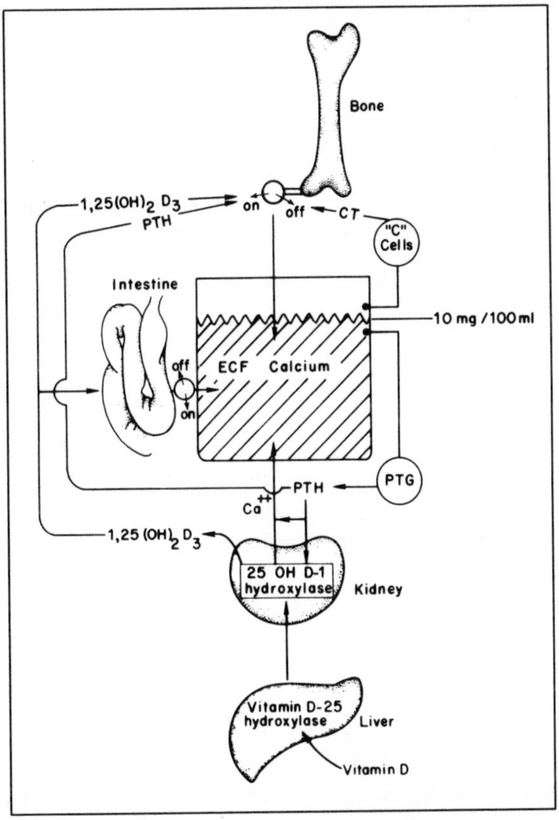

Fig. 3. Diagrammatic representation of the regulation of plasma calcium concentration by the vitamin D and parathyroid hormone endocrine system. Under conditions of hypocalcemia the parathyroid gland is stimulated to secrete parathyroid hormone. Parathyroid hormone accumulates in kidney and bone. In the kidney the parathyroid hormone increases calcium reabsorption, causes a phosphate diuresis (not shown) and stimulates production of 1,25-$(OH)_2D_3$. The 1,25-$(OH)_2D_3$ proceeds to the intestine where it stimulates intestinal calcium absorption and progresses to bone where, together with the parathyroid hormone, it mobilizes calcium from the bone fluid compartment. This results in an elevation of serum calcium to levels which suppress parathyroid hormone secretion.

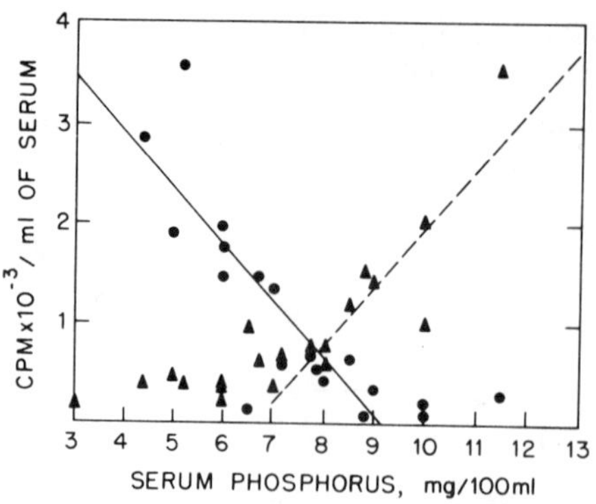

Fig. 4. The relationship of serum phosphorus concentration to production of 24,25-$(OH)_2D_3$ ▲ and 1,25-$(OH)_2D_3$ • from 25-OH-D_3 in thyroparathyroidectomized rats. Rats were thyroparathyroidectomized and placed on a variety of diets varying in phosphorus concentration, which adjusted serum phosphorus concentration to the indicated levels. After 3 days of adjustment, the *in vivo* production of metabolites from radioactive 25-OH-D_3 was assessed. Note that normal serum phosphorus concentration in young growing rats is about 9 mg/100 ml and in the thyroparathyroidectomized state these animals produce little or no 1,25-$(OH)_2D_3$. However, if they are made hypophosphatemic, 1,25-$(OH)_2D_3$ is produced even in the absence of parathyroid glands.

serum calcium concentration and in response to hypocalcemia stimulate secretion of the peptide hormone, parathyroid hormone (64). The parathyroid hormone then has been shown to proceed to the kidney, bone and liver (65, 66). In the kidney besides causing a phosphate diuresis (67) and improving renal reabsorption of calcium (68) it stimulates production of 1,25-(OH)-$_2D_3$. The 1,25-$(OH)_2D_3$ then proceeds to the intestine where it stimulates intestinal calcium transport without further aid from the parathyroid hormone (69). At the bone site both parathyroid hormone and 1,25-$(OH)_2D_3$ are required for the mobilization of calcium from bone. It is likely that the renal reabsorption of calcium in the distal tubule which is under parathyroid hormone regulation is also under regulation by both parathyroid hormone and 1,25-$(OH)_2D_3$. As will be discussed by Dr. *Massry,* it is clear that in the absence of 1,25-$(OH)_2D_3$ the skeleton becomes resistant to parathyroid hormone stimulation (70). By the same token the absence of parathyroid hormone markedly diminshes if not eliminates the response of bone to 1,25-$(OH)_2D_3$ (69).

In addition to regulation by serum calcium through the parathyroid glands, it has also been demonstrated that serum phosphorus concentration plays a regulatory role in 1,25-$(OH)_2D_3$ accumulation in plasma and target tissues (51). Furthermore, there has been direct evidence that phosphate deprivation stimulates 25-OH-D_3-1-hydroxylase (71). Thus if the calcium control mechanism is

Table I. Serum levels of 1,25-$(OH)_2D_3$ in chronic renal disease

	pg/ml
Normal	36 ± 2
Anephric	0
GFR <1	0
GFR 50	58, 44
GFR 20–40	24 ± 4

All pateints have intact parathyroid glands.

removed by thyroparathyroidectomy a regulation of serum levels of 1,25-$(OH)_2D_3$ by serum phosphorus concentration can be readily demonstrated (fig. 4) (51). Thus the need for phosphate can stimulate production of 1,25-$(OH)_2D_3$ even in the absence of the parathyroid glands. Therefore, it is quite clear that in addition to the parathyroid hormone, phosphate exerts a strong regulatory role on 1,25-$(OH)_2D_3$production. It is well known that phosphate deprivation stimulates intestinal calcium absorption (72, 73) and mobilization of calcium from bone (74). However, it is clear that in addition to stimulating the 1-hydroxylase, phosphate deprivation further stimulates intestinal transport of calcium and possibly mobilization of calcium from bone (17). The mechanism of this stimulation remains undefined at the present time. Nevertheless, the regulation of 1,25-$(OH)_2D_3$ production by phosphate is an important aspect of management of patients suffering from chronic renal failure.

Hyperphosphatemia is, of course, a major byproduct which results from phosphate retention as a result of loss of nephrons. It is indeed possible that the rising serum phosphate concentration may suppress the remaining 1-hydroxylase in the diseased kidneys giving a premature shut off of 1,25-$(OH)_2D_3$ production (75). In fact, a preliminary investigation of serum levels of 1,25-$(OH)_2D_3$ as related to glomerular filtration rate has revealed that at approximately a GFR of 20 ml/min, serum levels of 1,25-$(OH)_2D_3$ diminish (table I) (*Hamstra, Slatopolsky and DeLuca,* unpublished results). Thus an important component of renal osteodystrophy is a failure to manufacture the required amounts of 1,25-$(OH)_2D_3$ bringing about hypocalcemia and stimulated parathyroid hormone secretion. When there is a complete lack of 1,25-$(OH)_2D_3$, skeletal resistance to parathyroid hormone and osteomalacia results (76). It is important to note that as there is progressive renal failure and progressive diminution of renal mass it is likely that the kidneys fail to synthesize sufficient amounts of 1,25-$(OH)_2D_3$ to meet the calcium requirement of the individual. However, there is nevertheless some 1,25-$(OH)_2D_3$ made at this stage which permits the excessively secreted parathyroid hormone to mobilize skeletal calcium giving rise to the osteitis fibrosa and probably osteosclerosis. Resistance to the parathyroid hormone does not develop until late in the disease which corresponds to the lack of circulating 1,25-$(OH)_2D_3$. Experiments are now in progress together with Dr. *Slatopolsky* of Washington University, St. Louis, Mo. and Dr. *Simpson* of Madison to compare renal disease type with serum circulating levels of 1,25-$(OH)_2D_3$ and within each type to compare the glomerular filtration rate versus serum 1,25-$(OH)_2D_3$ values. Unfortunately, that study is not yet complete.

The ability of the intestine to adjust its calcium absorption to meet the organismal requirements for calcium is an important factor which prevents loss of skeletal stores of calcium to support serum calcium concentration and prevent osteoporosis. From the above discussion it is evident that upon prolonged deprivation of calcium or where there is a continual need for calcium there is a continual excitation of the parathyroid glands giving rise to parathyroid hormone. The parathyroid hormone then stimulates production of 1,25-$(OH)_2D_3$ which in turn stimulates the intestine to a higher efficiency of calcium absorption (17, 18, 48). It is, therefore, clear that the ability of the intestine to adapt to dietary calcium needs involves the 1,25-$(OH)_2D_3$ endocrine system (17, 18). There has been some argument that the parathyroid glands are not involved in the ability of the intestine to adapt to calcium requirement inasmuch as parathyroidectomized animals still show evidence of adaptation (77, 78). Unfortunately these experiments suffer from an inability to be certain that all of the parathyroid tissue has been removed giving rise to the very clear possibility that remaining parathyroid tissue does adapt to the need for calcium. Recently we have been able to demonstrate that the administration of a constant exogenous source of 1,25-$(OH)_2D_3$ will prevent the organism from adjusting their intestinal calcium absorption to dietary calcium levels (fig. 5) (17, 18). Furthermore, we can demonstrate that animals which are parathyroidectomized and then given an exogenous but constant source of parathyroid hormone (fig. 5) also lose the ability to adjust their intestine to the dietary calcium level, demonstrating clearly that the ability to adjust intestinal calcium absorption to dietary clacium is the province of parathyroid hormone-1,25-$(OH)_2D_3$ endocrine systems. It seems that a contributing factor to the development of osteoporosis is an inability to adjust the intestinal calcium absorption to meet the calcium needs of the individual giving rise to a reliance upon skeletal calcium for support of serum calcium concentration. A continuing raid on skeletal calcium over a period of many years gives rise to loss of bone mass or osteoporosis. The loss of bone mass in postmenopausal osteoporosis is also at least in part the result of an increased sensitivity of the skeleton in the absence of the sex hormones to factors which cause mobilization of calcium or bone resorption (79–81). Nevertheless, a clear understanding of this disease may result from an understanding of the vitamin D endocrine system and its relationship to calcium metabolism.

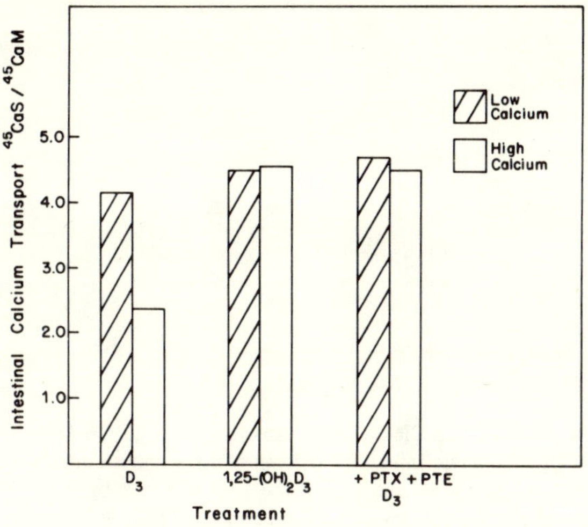

Fig. 5. The adaptation of intestinal calcium transport to low dietary calcium and its elimination by the administration of either exogenous 1,25-$(OH)_2D_3$ or parathyroid hormone. Rats were fed either a low calcium diet or a high calcium diet with normal levels of phosphorus for a period of 2 weeks. At this time they were either given oral vehicle or 250 ng of vitamin D_3 daily or 25 ng daily of 1,25-$(OH)_2D_3$. In the case of one group they were thyroparathyroidectomized and in addition to receiving the 250 ng of vitamin D_3 daily they received an injection of parathyroid hormone every 6 h for a period of 7 days. At the end of 7 days of treatment the animals were sacrificed and their intestinal calcium transport determined. Notice that exogenous parathyroid hormone or exogenous 1,25-$(OH)_2D_3$ eliminated the ability of the intestine to adapt to low dietary calcium levels.

The vitamin D endocrine system is now known to be regulated by other endocrine systems. In particular during the egg-laying cycle of birds, it is now known that the sex hormones, namely estradiol, testosterone and progesterone, interact to stimulate production of 1,25-$(OH)_2D_3$ (79, 80; *Tanaka and DeLuca,* unpublished). This can best be illustrated as shown in figure 6 which shows that the injection of estradiol to mature male Japanese quail brings about a suppression of 24-hydroxylase and a marked stimulation of the 1-hydroxylase in the kidneys followed by an elevation of serum calcium concentration. More recently it has been possible to show that in castrate male birds, estradiol, testosterone, and progesterone given individually will not stimulate the 1-hydroxylase but when they are given in the correct combination, concentration, and ratio there is a phenomenal stimulation of the 1-hydroxylase. It seems likely that at the time medullary bone is resorbed in birds in response to the need for calcium for egg shell production, there is a marked stimulation of the 1-hydroxylase, a subse-

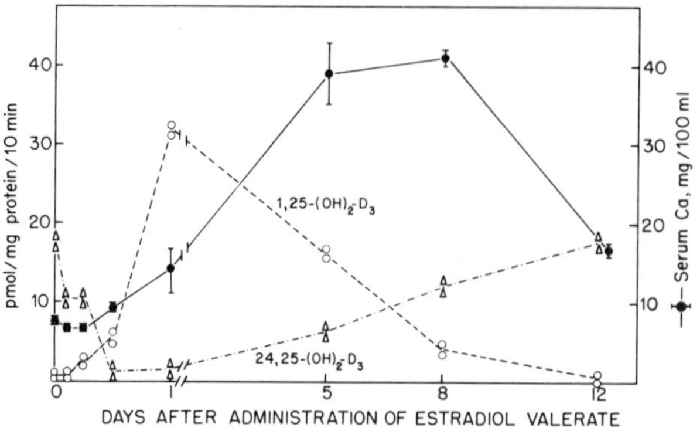

Fig. 6. The response of the renal 25-OH-D_3-24-hydroxylase and 25-OH-D_3-1-hydroxylase of mature Japanese quail to a single injection of 5 mg of estradiol valerate. Note the suppression of the 24-hydroxylase and the stimulation of the 1-hydroxylase by the estradiol treatment. This is followed by an elevation of serum calcium concentration. Measurement of serum levels of 1,25-$(OH)_2D_3$ (not shown) also demonstrate the elevation of serum 1,25-$(OH)_2D_3$ levels in parallel with the elevation of the 1-hydroxylase.

quent elevation of plasma 1,25-$(OH)_2D_3$ levels, a disappearance of medullary bone and a deposit of calcium carbonate in egg shell. Therefore, the sex hormones can regulate the vitamin D endocrine system for these extraordinary needs of calcium and this phenomenon may have its implications in postmenopausal metabolic bone disease.

It should be mentioned that work by *Schneider et al.* (82, 83) has shown that diabetes induced by streptozotocin or alloxan causes a diminution of calcium absorption which can be corrected by the administration of 1,25-$(OH)_2D_3$. Furthermore, low circulating levels of 1,25-$(OH)_2D_3$ can be found in the diabetic rat and can be restored at least in part by the administration of insulin (84). Work has been carried out by *Spanos and MacIntyre* (85) to demonstrate that prolactin will also stimulate the 25-OH-D_3-1-hydroxylase. These results demonstrate that other endocrine systems which affect calcium metabolism may well do so by regulating production of 1,25-$(OH)_2D_3$.

In conclusion it appears that the vitamin D endocrine system may be the central endocrine system in the regulation of calcium and phosphorus metabolism and that many factors which influence calcium metabolism may do so by influencing the vitamin D endocrine system located in the kidney. Thus in renal failure or renal disease a disturbance of calcium metabolism under a variety of circumstances is clearly understandable and it is well known that a large number of patients suffering from renal osteodystrophy respond to physiologic doses of

$1,25\text{-}(OH)_2D_3$ or its analog $1\alpha\text{-}OH\text{-}D_3$ (86–88). Thus the vitamin D endocrine system may assume central importance in the regulation of calcium and possibly phosphorus metabolism, previously unappreciated in the absence of current information available on vitamin D metabolism.

Summary

During the past decade, it has been clearly established that the kidney is an endocrine organ which converts 25-hydroxyvitamin D to the active hormone 1,25-dihydroxyvitamin D. The kidney and intestine are the sites of an alternative hydroxylation which places an 24 hydroxyl in the *R* position on either 25-hydroxyvitamin D or 1,25-dihydroxyvitamin D. The production of 1,25-dihydroxyvitamin D is stimulated by hypocalcemia through the parathyroid gland, hypophosphatemia, the sex hormones and the peptide hormones insulin and prolactin. This vitamin D based endocrine system is of central importance in regulating calcium and phosphorus metabolism and its disturbances result in a variety of pathological conditions.

References

1 *DeLuca, H. and Schnoes, H.:* Metabolism and mechanism of action of vitamin D. A. Rev. Biochem. *45:* 631–666 (1976).

2 *Kodicek, E.:* The story of vitamin D, from vitamin to hormone. Lancet *i:* 325–329 (1974).

3 *DeLuca, H.:* Recent advances in our understanding of the vitamin D endocrine system. J. Lab. clin. Med. *87:* 7–26 (1976).

4 *Norman, A. and Henry, H.:* 1,25-Dihydroxyvitamin D_3. A hormonally active form of vitamin D_3. Recent Prog. Horm. Res. *30:* 431–480 (1974).

5 *DeLuca, H.:* Mechanism of action and metabolic fate of vitamin D. Vitams Horm. *25:* 315–367 (1967).

6 *Omdahl, J. and DeLuca, H.:* Regulation of vitamin D metabolism and function. Physiol. Rev. *53:* 327–372 (1973).

7 *Lamm, M. and Neuman, W.F.:* On the role of vitamin D in calcification, Archs Path. *66:* 204–209 (1958).

8 *Neuman, W. and Neuman, M.:* The chemical dynamics of bone mineral (University of Chicago Press, Chicago 1958).

9 *Reed, C.; Struck, H., and Steck, I.:* Vitamin D (University of Chicago Press, Chicago 1939).

10 *Nicolaysen, R. and Eeg-Larsen, N.:* The biochemistry and physiology of vitamin D. Vitams Horm. *11:* 29–60 (1953).

11 *Wasserman, R.:* The transfer of calcium and strontium across biological membranes (Academic Press, New York 1963).

12 *Harrison, H. and Harrison, H.:* Intestinal transport of phosphate: action of vitamin D, calcium and potassium. Am. J. Physiol. *201:* 1007–1012 (1961).

13 *Chen, T.; Castillo, L.; Korycka-Dahl, M., and DeLuca, H.:* Role of vitamin D metabolites in phosphate transport of rat intestine. J. Nutr. *104:* 1056–1060 (1974).

14 *Steele, T.; Engle, J.; Tanaka, Y.; Lorenc, R.; Dudgeon, K., and DeLuca, H.:* Phosphatemic action of 1,25-dihydroxyvitamin D_3. Am. J. Physiol. *229:* 489–495 (1975).

15 *Costanzo, L.; Sheehe, P., and Weiner, I.:* Renal actions of vitamin D in D-deficient rats. Am. J. Physiol *226:* 1490–1495 (1974).

16 *Nicolaysen, R.; Eeg-Larsen, N., and Malm, O.:* Physiology of calcium metabolism. Physiol. Rev. *33:* 424–444 (1953).

17 *Ribovich, M. and DeLuca, H.:* The influence of dietary calcium and phosphorus on intestinal calcium transport in rats given vitamin D metabolites. Archs Biochem. Biophys. *170:* 529–535 (1975).

18 *Ribovich, M. and DeLuca, H.:* Intestinal calcium transport: parathyroid hormone and adaptation to dietary calcium. Archs Biochem. Biophys. *175:* 256–261 (1976).

19 *DeLuca, H.; Blunt, J., and Rikkers, H.:* VII. Biogenesis; in The vitamins (Academic Press, New York 1971).

20 *Esvelt, R.; DeLuca, H., and Schnoes, H.:* The production of vitamin D_3 in ultraviolet irradiated rat skin. Abstract. Proc. 6th Parathyroid Conf., 1977.

21 *Mellanby, E.:* An experimental investigation on rickets. Lancet *1:* 407–412 (1919).

22 *Ponchon, G. and DeLuca, H.:* The role of the liver in the metabolism of vitamin D. J. clin. Invest. *48:* 1273–1279 (1969).

23 *Ponchon, G.; Kennan, A.L., and DeLuca, H.F.:* Activation of vitamin D by the liver. J. clin. Invest. *48:* 2032–2037 (1969).

24 *Horsting, M. and DeLuca, H.F.: In vitro* production of 25-hydroxycholecalciferol. Biochem. biophys. Res. Commun. *36:* 251–256 (1969).

25 *Blunt, J.W.; DeLuca, H.F., and Schnoes, H.K.:* 25-Hydroxycholecalciferol. A biologically active metabolite of vitamin D_3. Biochemistry *7:* 3317–3322 (1968).

26 *Bhattacharyya, M. and DeLuca, H.F.:* Subcellular location of rat liver calciferol-25-hydroxylase. Archs Biochem. Biophys. *160:* 58–62 (1974).

27 *Bhattacharyya, M.H. and DeLuca, H.F.:* The regulation of rat liver calciferol-25-hydroxylase. J. biol. Chem. *248:* 2969–2973 (1973).

28 *Haddad, J.G. and Stamp, T.C.B.:* Circulating 25-hydroxyvitamin D in man. Am. J. Med. *57:* 57–62 (1974).

29 *Tanaka, Y.; Frank, H., and DeLuca, H.F.:* Biological activity of 1,25-dihydroxyvitamin D_3 in the rat. Endocrinology *92:* 417–422 (1973).

30 *Imawari, M.; Kida, K., and Goodman, D.S.:* The transport of vitamin D and its 25-hydroxy metabolite in human plasma, J. clin. Invest. *58:* 514–523 (1976).

31 *Botham, K.M.; Ghazarian, J.G.; Kream, B.E., and DeLuca, H.F.:* Isolation of a potent inhibitor of 25-hydroxyvitamin D_3-1-hydroxylase from rat serum. Biochemistry *15:* 2130–2135 (1976).

32 *Haddad, J.G. and Walgate, J.:* 25-Hydroxyvitamin D transport in human plasma. J. biol. Chem. *251:* 4803–4809 (1976).

33 *Gray, R.W.; Omdahl, J.L.; Ghazarian, J.G., and DeLuca, H.F.:* 25-Hydroxycholecalciferol-1-hydroxylase: subcellular location and properties. J. biol. Chem. *247:* 7528–7532 (1972).

34 *Holick, M.F.; Schnoes, H.K.; DeLuca, H.F.; Suda, T., and Cousins, R.J.:* Isolation and identification of 1,25-dihydroxycholecalciferol. A metabolite of vitamin D active in intestine. Biochemistry *10:* 2799–2804 (1971).

35 *Fraser, D.R. and Kodicek, E.:* Unique biosynthesis by kidney of a biologically active vitamin D metabolite. Nature, Lond. *228:* 764–766 (1970).

36 *Gray, R.; Boyle, I., and DeLuca, H.F.:* Vitamin D metabolism: the role of kidney tissue. Science *172:* 1232–1234 (1971).

37 *Ghazarian, J.G.; Schnoes, H.K., and DeLuca, H.F.:* Mechanism of 25-hydroxycholecal-

ciferol 1α-hydroxylation. Incorporation of oxygen-18 into the 1α position of 25-hydroxycholecalciferol. Biochemistry *12:* 2555–2558 (1973).

38 *Ghazarian, J.G.; Jefcoate, C.R.; Knutson, J.C.; Orme-Johnson, W.H., and DeLuca, H.F.:* Mitochondrial cytochrom P_{450}: a component of chick kidney 25-hydroxycholecalciferol-1α-hydroxylase. J. biol. Chem. *249:* 3026–3033 (1974).

39 *Pedersen, J.I.; Ghazarian, J.G.; Orme-Johnson, N.R., and DeLuca, H.F.:* Isolation of chick renal mitochondrial ferredoxin active in the 25-hydroxyvitamin D_3-1α-hydroxylase system. J. biol. Chem. *251:* 3933–3941 (1976).

40 *Boyle, I.T.; Miravet, L.; Gray, R.W.; Holick, M.F., and DeLuca, H.F.:* The response of intestinal calcium transport to 25-hydroxy and 1,25-dihydroxy vitamin D in nephrectomized rats. Endocrinology *90:* 605–608 (1972).

41 *Holick, M.F.; Garabedian, M., and DeLuca, H.F.:* 1,25-Dihydroxycholecalciferol: metabolite of vitamin D_3 active on bone in anephric rats. Science *176:* 1146–1147 (1972).

42 *Wong, R.G.; Norman, A.W.; Reddy, C.R., and Coburn, J.W.:* Biologic effects of 1,25-dihydroxycholecalciferol (a highly active vitamin D metabolite) in acutely uremic rats. J. clin. Invest. *51:* 1287–1291 (1972).

43 *Frolik, C.A. and DeLuca, H.F.:* 1,25-Dihydroxycholecalciferol: the metabolite of vitamin D responsible for increased intestinal calcium transport. Archs Biochem. Biophys. *147:* 143–147 (1971).

44 *Frolik, C.A. and DeLuca, H.F.:* Metabolism of 1,25-dihydroxycholecalciferol in the rat. J. clin. Invest. *51:* 2900–2906 (1972).

45 *Tsai, H.C.; Wong, R.G., and Norman, A.W.:* Studies on calciferol metabolism. IV. Subcellular localization of 1,25-dihydroxy-vitamin D_3 in intestinal mucosa and correlation with increased calcium transport. J. biol. Chem. *247:* 5511–5519 (1972).

46 *Kumar, R.; Harnden, D., and DeLuca, H.F.:* Metabolism of 1,25-dihydroxyvitamin D_3: evidence for side-chain oxidation. Biochemistry *15:* 2420–2423 (1976).

47 *DeLuca, H.F.:* Vitamin D: the vitamin and the hormone. Fed. Proc. Fed. Am. Socs exp. Biol. *33:* 2211–2219 (1974).

48 *Boyle, I.T.; Gray, R.W., and DeLuca, H.F.:* Regulation by calcium of *in vivo* synthesis of 1,25-dihydroxycholecalciferol and 21,25-dihydroxycholecalciferol. Proc. natn. Acad. Sci. USA *68:* 2131–2134 (1971).

49 *Boyle, I.T.; Gray, R.W.; Omdahl, J.L., and DeLuca, H.F.:* Calcium control of the *in vivo* biosynthesis of 1,25-dihydroxyvitamin D_3: Nicolaysen's endogenous factor; in Endocrinology 1971 (Heinemann, London 1972).

50 *Omdahl, J.L.; Gray, R.W.; Boyle, I.T.; Knutson, J., and DeLuca, H.F.:* Regulation of metabolism of 25-hydroxycholecalciferol by kidney tissue *in vitro* by dietary calcium. Nature new Biol. *237:* 63–64 (1972).

51 *Tanaka, Y. and DeLuca, H.F.:* The control of 25-hydroxyvitamin D metabolism by inorganic phosphorus. Archs Biochem. Biophys. *154:* 566–574 (1973).

52 *Garabedian, M.; Holick, M.F.; DeLuca, H.F., and Boyle, I.T.:* Control of 25-hydroxycholecalciferol metabolism by the parathyroid glands. Proc. natn. Acad. Sci. USA *69:* 1673–1676 (1972).

53 *Holick, M.F.; Schnoes, H.K.; DeLuca, H.F.; Gray, R.W.; Boyle, I.T., and Suda, T.:* Isolation and identification of 24,25-dihydroxycholecalciferol: a metabolite of vitamin D_3 made in the kidney. Biochemistry *11:* 4251–4255 (1972).

54 *Knutson, J.C. and DeLuca, H.F.:* 25-Hydroxyvitamin D_3-24-hydroxylase: subcellular location and properties. Biochemistry *13:* 1543–1548 (1974).

55 *Tanaka, Y. and DeLuca, H.F.:* Stimulation of 24,25-dihydroxyvitamin D_3 production by 1,25-dihydroxyvitamin D_3. Science *183:* 1198–1200 (1974).

56 *Tanaka, Y.; Lorenc, R.S., and DeLuca, H.F.:* The role of 1,25-dihydroxyvitamin D_3 and parathyroid hormone in the regulation of chick renal 25-hydroxyvitamin D_3-24-hydroxylase. Archs Biochem. Biophys. *171:* 521–526 (1975).

57 *DeLuca, H.F.:* Minireview: regulation of vitamin D metabolism. Life Sci. *17:* 1351–1358 (1975).

58 *DeLuca, H.F. and Schnoes, H.K.:* Recent advances in our understanding of the metabolism and mechanism of action of 1,25-dihydroxyvitamin D_3. Proc. 6th Parathyroid Conf. (Excerpta Medica, Amsterdam, in press 1977).

59 *Tanaka, Y.; Castillo, L.; DeLuca, H.F., and Ikekawa, N.:* The 24-hydroxylation of 1,25-dihydroxyvitamin D_3. J. biol. Chem. *252:* 1421–1424 (1977).

60 *Tanaka, Y.; DeLuca, H.F.; Ikekawa, N.; Morisaki, M., and Koizumi, N.:* Determination of stereochemical configuration of the 24-hydroxyl group of 24,25-dihydroxyvitamin D_3 and its biological importance. Archs Biochem. Biophys. *170:* 620–626 (1975).

61 *Holick, M.F.; Baxter, L.A.; Schraufrogel, P.K.; Tavela, T.E., and DeLuca, H.F.:* Metabolism and biological activity of 24,25-dihydroxyvitamin D_3 in the chick, J. biol. Chem. *251:* 397–402 (1976).

62 *Henry, H.L.; Midgett, R.J., and Norman, A.W.:* Regulation of 25-hydroxyvitamin D_3-1-hydroxylase *in vivo.* J. biol. Chem. *249:* 7584–7592 (1974).

63 *Fraser, D.R. and Kodicek, E.:* Regulation of 25-dydroxycholecalciferol-1-hydroxylase activity in kidney by parathyroid hormone. Nature new Biol. *241:* 163–166 (1973).

64 *Potts, J.T., jr.:* Introductory remarks; in Parathyroid hormone and thyrocalcitonin (calcitonin) (Excerpta Medica, Amsterdam 1968).

65 *Zull, J.E. and Repke, D.W.:* The tissue localization of tritiated parathyroid hormone in thyroparathyroidectomized rats. J. biol. Chem. *247:* 2195–2199 (1972).

66 *Neuman, M.W.; Neuman, W.F., and Lane, K.:* The metabolism of labeled parathyroid hormone. VI. Effects of vitamin D status. Calcif. Tiss. Res. *18:* 289–295 (1975).

67 *Rasmussen, H.:* Parathyroid hormone, calcitonin and the calciferols; in Textbook of endocrinology (Saunders, Philadelphia 1974).

68 *Kleeman, C.R.; Bernstein, D.; Rockney, R.; Dowling, J.T., and Maxwell, M.H.:* Studies on the renal clearance of diffusible calcium and the role of the parathyroid glands in its regulation; in The parathyroids (Thomas, Springfield 1961).

69 *Garabedian, M.; Tanaka, Y.; Holick, M.F., and DeLuca, H.F.:* Response of intestinal calcium transport and bone calcium mobilization to 1,25-dihydroxyvitamin D_3 in thyroparathyroidectomized rats. Endocrinology *94:* 1022–1027 (1974).

70 *Massry, S.G.; Coburn, J.W.; Lee, D.B.N.; Jowsey, J., and Kleeman, C.R.:* Skeletal resistance to parathyroid hormone in renal failure. Ann. intern. Med. *78:* 357–364 (1973).

71 *Baxter, L.A. and DeLuca, H.F.:* Stimulation of 25-hydroxyvitamin D_3-1α-hydroxylase by phosphate depletion. J. biol. Chem. *251:* 3158–3161 (1976).

72 *Tanaka, Y.; Frank, H., and DeLuca, H.F.:* Intestinal calcium transport: stimulation by low phosphorus diets. Science *181:* 564–566 (1973).

73 *Morrissey, R.L. and Wasserman, R.H.:* Calcium absorption and calcium-binding protein in chicks on differing calcium and phosphorus intakes. Am. J. Physiol. *220:* 1509–1515 (1971).

74 *Baylink, D.; Wergedal, J., and Stauffer, M.:* Formation, mineralization, and resorption of bone in hypophosphatemic rats. J. clin. Invest. *50:* 2519–2530 (1971).

75 *DeLuca, H.F.:* The role of the kidney in the metabolism of vitamin D; in Proc. 5th Int. Congr. Nephrology, Mexico 1972 (Karger, Basel 1974).

76 *DeLuca, H.F. and Avioli, L.V.:* Renal osteodystrophy; in Renal disease (Blackwell, Oxford, in press 1977).

77 *Kimberg, D.V.; Schachter, D., and Schenker, H.:* Active transport of calcium by intestine: effects of dietary calcium. Am. J. Physiol. *200:* 1256–1262 (1961).
78 *Favus, M.J.; Walling, M.W., and Kimberg, D.V.:* Effects of dietary calcium restriction and chronic thyroparathyroidectomy on the metabolism of [^{3}H]-25-hydroxy-vitamin D_3 and the active transport of calcium by rat intestine. J. clin. Invest. *53:* 1139–1148 (1974).
79 *Tanaka, Y.; Castillo, L., and DeLuca, H.F.:* Control of the renal vitamin D hydroxylases in birds by the sex hormones. Proc. natn. Acad. Sci. USA *73:* 2701–2705 (1976).
80 *Castillo, L.; Tanaka, Y.; DeLuca, H.F., and Sunde, M.L.:* The stimulation of 25-hydroxyvitamin D_3-1α-hydroxylase by estrogen. Archs Biochem. Biophys. *179:* 211–217 (1977).
81 *Kenny, A.D.:* Vitamin D metabolism: physiological regulation in egg-laying Japanese quail. Am. J. Physiol. *230:* 1609–1615 (1976).
82 *Schneider, L.E.; Wilson, H.D., and Schedl, H.P.:* Effects of alloxan diabetes on duodenal calcium-binding protein in the rat. Am. J. Physiol. *227:* 832–838 (1974).
83 *Schneider, L.E.; Omdahl, J.L., and Schedl, H.P.:* Effects of vitamin D and its metabolites on calcium transport in the diabetic rat. Endocrinology *99:* 793–799 (1976).
84 *Haussler, M.R.:* Biochemical, physiologic, and clinical applications of radioligand receptor assay for 1,25-dihydroxyvitamin D; in Vitamin D: biochemical, chemical and clinical aspects related to calcium metabolism (de Gruyter, Berlin in press 1977).
85 *Spanos, E. and MacIntyre, I.:* Vitamin D and the pituitary, Lancet *i:* 840–841 (1977).
86 *Silverberg, D.S.; Bettcher, K.B.; Dossetor, J.B.; Overton, T.R.; Holick, M.F., and DeLuca, H.F.:* Effect on 1,25-dihydroxycholecalciferol in renal osteodystrophy. Can. med. Ass. J. *112:* 190–195 (1975).
87 *Brickman, A.S.; Sherrard, D.J.; Jowsey, J.; Singer, F.R.; Baylink, D.J.; Maloney, N.; Massry, S.G.; Norman, A.W., and Coburn, J.W.:* 1,25-Dihydroxycholecalciferol. Effect on skeletal lesions and plasma parathyroid hormone levels in uremic osteodystrophy. Archs intern. Med. *134:* 883–888 (1974).
88 *Chan, J.C.M.; Oldham, S.B.; Holick, M.F., and DeLuca, H.F.:* 1α-Hydroxyvitamin D_3 in chronic renal failure. A potent analogue of the kidney hormone, 1,25-dihydroxy-cholecalciferol. J. Am. med. Ass. *234:* 47–52 (1975).

Dr. *H.F. DeLuca*, Department of Biochemistry, College of Agricultural and Life Sciences, University of Wisconsin-Madison, *Madison, WI 53706* (USA)

Contr. Nephrol., vol. 13, pp. 96–103 (Karger, Basel 1978)

Effect of Vitamin D Metabolites on Calcium and Phosphate Metabolism[1]

H. Fleisch, J.-P. Bonjour, R. Rizzoli and K. Hugi
Department of Pathophysiology, University of Berne, Berne

In recent years it has been established that 1,25-dihydroxyvitamin D_3 (1,25-$(OH)_2D_3$) is the main active metabolite of vitamin D_3. 1,25-$(OH)_2D_3$ has been shown to correct both the decreased intestinal absorption of calcium and phosphate (Pi) as well as the deficient mineralization of bone in rachitic animals (1–3). However, few data are available as yet on its effect in animals with a normal vitamin D status, and in animals which are deficient in 1,25-$(OH)_2D_3$ but with an otherwise normal D status. The effect of 1,25-$(OH)_2D_3$ in normal animals is of interest, since an interplay between the effect of 1,25-$(OH)_2D_3$ and other D metabolites could lead to an overall increased calcium retention in bone. Such an effect could be of use in the therapy of clinical bone disease of man. Furthermore the study of the effect of a specific deficiency in 1,25-$(OH)_2D_3$ will give us a better understanding of the physiological role of this metabolite *in vivo*.

This paper will report results on the effect of 1,25-$(OH)_2D_3$ on (a) calcium metabolism in normal rats, and (b) calcium metabolism, (c) renal handling of calcium, (d) intestinal absorption of Pi, and (e) renal handling of Pi in 1,25-$(OH)_2D_3$-deficient rats.

[1] This work was supported by the Swiss National Science Foundation (3.725.76), by F. Hoffmann-La Roche & Co., Basel, Switzerland, and by the Procter and Gamble Company, USA.

Effect of Vitamin D Metabolites in Normal Rats (4)

The fact that the metabolites of vitamin D_3 are able to increase the intestinal absorption of calcium suggested that they might also lead to an increased retention of this ion. Indeed it is known that if rats are fed increasing amounts of calcium, they will also retain in their skeleton increased amounts of this ion. This positive balance is due to a decrease in bone resorption in response to the higher intestinal absorption, while calcium deposition in bone is not altered (5–7). We therefore investigated the effect of various vitamin D metabolites on the different parameters of calcium metabolism measured by means of ^{45}Ca kinetics. The animals were all pair-fed and received vitamin D_3 daily *per os.* The various calcium fluxes were measured at the age of 61 ± 1 days by the technique of *Aubert and Milhaud* (8).

It appeared that 1,25-$(OH)_2D_3$ given at a dose of 2 X 13 pmol/day i.p. caused an increase in net absorption of calcium but did not affect bone formation and bone resorption. The balance was not significantly altered either, although a tendency to an increase occurred. The increased intestinal absorption was compensated by an increase in urinary calcium. These results were similar to those obtained previously with biosynthetic 1,25-$(OH)_2D_3$ (9). A larger dose of 1,25-$(OH)_2D_3$, 2 X 65 pmol/day, increased even more intestinal absorption and calciuria, but now actually decreased bone formation and resorption as well as calcium retention.

Both the R and S isomers of 24,25-$(OH)_2D_3$ given at 2 X 13, 2 X 130 and 2 X 650 pmol/day i.p. had no significant effect on any of the fluxes measured. The R and S isomers of 1,24,25-$(OH)_3D_3$ given at 2 X 13 pmol/day i.p. had no significant effect either. However, at a daily dose of 2 X 130 pmol i.p. 1,24R,25-$(OH)_3D_3$ produced an increase in the intestinal absorption similar to that obtained with one-tenth the dose of 1,25-$(OH)_2D_3$. As with 1,25-$(OH)_2D_3$ this effect was associated with a marked elevation of urinary calcium, no change in bone formation or resorption and no significant change in calcium balance. Interestingly with the S isomer, given at this dose, no effect on calcium absorption could be detected but the rise in urinary calcium was present. Finally 2 X 130 pmol/day of 1,25-$(OCOCH_3)_2D_3$ had no effect on any of the parameters measured.

Thus none of the various vitamin D metabolites or analogues tested could induce an increase in osseous calcium retention, despite the marked effect some of them had on intestinal calcium absorption. It seems that the decrease in bone resorption which normally accompanies an increased gut absorption of calcium does not occur when the intestinal effect is obtained by vitamin D metabolites, perhaps because of the intrinsic effect of the vitamin D metabolites to increase bone destruction (10, 11). In order to obtain an effect on balance one should envisage in future either an association of the metabolites with a compound inhibiting bone resorption, or the development of a new metabolite which has an intestinal effect but no action on bone resorption. However it must be stressed that these results were all obtained on normal growing animals. It cannot be ruled out at the present time that the effect could be different in old or diseased animals.

Role of 1,25-$(OH)_2D_3$ in Thyroparathyroidectomy (12)

It is known that thyroparathyroidectomy decreases the production of 1,25-$(OH)_2D_3$ (13, 14). The question arises to what extent the effect of parathyroid hormone (PTH) deprivation on calcium metabolism, namely the decrease in intestinal calcium absorption, in plasma calcium, in formation and bone destruction and in tubular absorption of calcium, is due respectively to the lack of PTH, and to the decrease of 1,25-$(OH)_2D_3$. Recent data (15) suggested that the effect of PTH on the intestinal absorption of calcium was due to the decrease in 1,25-$(OH)_2D_3$. We could confirm this suggestion. Indeed the effect of PTH on calcium absorption was abolished when the renal transformation of 25-OHD_3 into 1,25-$(OH)_2D_3$ was blocked by the diphosphonate disodium ethane-1-hydroxy-1,1-diphosphonate (EHDP) in rats (16). On the other hand EHDP had no effect on the action of 1,25-$(OH)_2D_3$. If the decrease in calcium absorption in thyroparathyroidectomized animals is due to the decrease in 1,25-$(OH)_2D_3$, the determination of the amount of the metabolite necessary to correct the intestinal absorption should be a measure of the decreased production after the ablation of the glands. This amount was found to be about 26 pmol/day. Interestingly this amount is the same necessary to correct the decrease in intestinal calcium absorption seen after administration of EHDP in doses which block the production of 1,25-$(OH)_2D_3$ *in vivo* (16). Thus it can be considered to be in the range of the physiological amount produced *in vivo*.

By administering to thyroparathyroidectomized rats the amount of 1,25-$(OH)_2D_3$ which just corrected the intestinal calcium absorption it was possible to determine how much of the other changes were due to 1,25-$(OH)_2D_3$ and PTH, respectively. It appeared that the metabolite increased to a certain extent both bone formation and bone destruction, without however fully restoring them to normal. The same is true for plasma calcium, although this parameter was more fully restored. Of interest is the fact that the D metabolite led to a significant increase in calciuria. This is probably due to the lack of the stimulating effect of PTH on renal tubular reabsorption. Similar results as those obtained with 1,25-$(OH)_2D_3$ were also found with 1,24,25-$(OH)_3D_3$ given however at a tenfold higher dose.

Thus part of the changes in calcium metabolism seen after thyroparathyroidectomy are due to the lack of 1,25-$(OH)_2D_3$. Clinically this suggests that patients with hypoparathyroidism could be treated with 1,25-$(OH)_2D_3$. This therapy should normalize blood calcium, but might lead to a hypercalciuria. These predictions have recently been found to be correct (17, 18).

Effect of 1,25-$(OH)_2D_3$ on the Renal Handling of Calcium (19)

The fact that 1,25-$(OH)_2D_3$ and 1,24,25-$(OH)_2D_3$ lead to an increase in calciuria raised the question whether this effect was a consequence of the hypercalcemia induced by the metabolites, or a direct renal action of the latter. In order to investigate this possibility the relationship between the urinary excretion of calcium per milliliter of glomerular filtrate and plasma calcium concentration was studied by clearance technique in thyroparathyroidectomized rats with or without 1,25-$(OH)_2D_3$ treatment and in rats receiving EHDP in an amount known to decrease the production of 1,25-$(OH)_2D_3$. Calcemia was varied between 1.5 and 3.5 m*M* by infusing calcium gluconate.

The administration of 2 × 13 pmol i.p. for 7 days of 1,25-$(OH)_2D_3$ to the thyroparathyroidectomized rats had no influence on the curve obtained, altering neither the 'threshold' nor the slope of the curve. The administration of EHDP had no influence either, nor did the administration of 1,25-$(OH)_2D_3$ in doses normalizing the intestinal absorption of calcium in the EHDP-treated rats.

Thus the kidney does not appear to alter its tubular handling of calcium in response to endogenous variations in 1,25-$(OH)_2D_3$ or in response to supplementation of 1,25-$(OH)_2D_3$ in physiological amounts.

Effect of 1,25-$(OH)_2D_3$ on Intestinal Pi Absorption (20)

It is known that in D-deficient rats impaired intestinal Pi absorption can be corrected by 1,25-$(OH)_2D_3$ (21–23). We have now investigated whether variations in the endogenous production of 1,25-$(OH)_2D_3$, when the D supply is normal, can also modulate Pi absorption in the gut. Three conditions known to alter 1,25-$(OH)_2D_3$ production, namely variation in Pi intake (24, 25), thyroparathyroidectomy (13, 14) and EHDP (16, 26, 27) were studied. In each, Pi absorption was measured by means of the *in situ* duodenal loop technique and by overall Pi balance in rats. The two techniques gave similar results in all circumstances.

It was found that lowering dietary Pi which is known to stimulate 1,25-$(OH)_2D_3$ production (24, 25), stimulated Pi absorption. Thyroparathyroidectomy on the other hand decreased it, an effect which was corrected by the administration of physiological amounts of either PTH or 1,25-$(OH)_2D_3$. Similarly, EHDP decreased the absorption, an effect which was again corrected by 1,25-$(OH)_2D_3$. Interestingly in thyroparathyroidectomized animals treated with EHDP, Pi absorption could be corrected by 1,25-$(OH)_2D_3$ but not by PTH, suggesting that as for calcium, the effect of PTH in intestinal ion transport is mediated by 1,25-$(OH)_2D_3$.

Thus it appears that variations in the endogenous production of 1,25-$(OH)_2D_3$ can modulate the rate of Pi absorption. However these changes are of relatively small importance when compared to the effect of variations in the dietary intake, so that in contrast with calcium absorption the intestinal modulation of Pi absorption can only be of minor importance in controlling Pi homeostasis.

Effect of 1,25-$(OH)_2D_3$ on the Renal Handling of Pi (28)

The final experiments were designed to investigate whether 1,25-$(OH)_2D_3$ had an effect on the renal handling of Pi. The effect of vitamin D and its metabolites on this handling is still a subject of controversy (29). While some investigators have described an increase in tubular reabsorption after administration of vitamin D, others have observed a decrease. Injection or infusion of 1,25-$(OH)_2D_3$ to dogs (30) or rats (31) has been found to depress acutely the fractional excretion of Pi. In the rat a large dose of the metabolite was necessary to obtain the effect. In the dog there was also a concomitant change in sodium excretion. Thus the physiological meaning of these effects can be questioned.

Our experiments were performed in intact and thyroparathyroidectomized rats, using clearance techniques. The fractional excretion of phosphate was measured at various plasma phosphate levels, which were obtained by infusing phosphate. Thus the excretion could be compared at a similar filtered load, which is a condition sine qua non for a meaningful comparison. 1,25-$(OH)_2D_3$ was given chronically at the physiological dose of 2 × 13 pmol/day for 7 days. In intact animals this treatment had no detectable effect. On the other hand the same dose decreased drastically the tubular reabsorption in thyroparathyroidectomized animals, restoring the curve to that seen in nonthyroparathyroidectomized controls. Under the same conditions 25-OHD_3 in doses of 2 × 13 pmol/day as well as 10 and 100 times higher doses had no effect. However, 2 × 13,000 pmol/day tended to correct the difference between normal and thyroparathyroidectomized rats. A correction is obtained with 2 × 130 pmol of 1,24,25-$(OH)_3D_3$ (R and S) whereas the same dose of 24R,25-$(OH)_2D_3$ had no effect. The effect of 1,25-$(OH)_2D_3$ was not secondary to an increase in sodium excretion, and was not accompanied by an increase in cAMP excretion.

These results suggest that a physiological change in the production of 1,25-$(OH)_2D_3$ can alter the renal handling of Pi, and that at least part of the effect of a lack of PTH is due to a decrease in 1,25-$(OH)_2D_3$ production.

Summary

Numerous studies have shown the effect of the vitamin D metabolites in vitamin D-deficient animals. However, little data are available on animals with a normal vitamin D status and in animals with a pure 1,25-$(OH)_2D_3$ deficiency. Thus the effect of administering physiological doses of 1,25-$(OH)_2D_3$ and other vitamin D metabolites and of modulating the endogenous production of 1,25-$(OH)_2D$ was investigated in the rat maintained on a normal vitamin D intake.

The administration of a physiological replacement dose of 1,25-$(OH)_2D_3$ (26 pmol/day i.p.) increases intestinal absorption and urinary excretion of calcium. Calcium balance is increased to a small but not significant extent. 1,24R,25-$(OH)_3D_3$ has a similar effect at a ten times higher dose. 24,25-$(OH)_2D_3$ and 1,25-$(OCOCH_3)_2D_3$ have no effects even at 500 and 10-fold higher doses, respectively.

In thyroparathyroidectomized animals it appears that the decrease in intestinal absorption of calcium is due entirely to the decrease in 1,25-$(OH)_2D$ production. Changes in plasma calcium are largely corrected by replacement with the vitamin D_3 metabolite. The change in bone turnover, however, is influenced only to a small degree. The change in renal handling of calcium is not altered at all.

Changing the production of 1,25-$(OH)_2D$ by altering dietary intake of Pi, parathyroidectomy, or administration of EHDP alters the intestinal absorption of phosphate. As in the case of calcium the effects of parathyroidectomy and EHDP treatment appear to be due entirely to the change in 1,25-$(OH)_2D$ production.

Finally, it was found that the increase in the net renal reabsorption of Pi observed after thyroparathyroidectomy can be completely reversed by administration of the lacking amount of 1,25-$(OH)_2D_3$. Physiological doses of this metabolite thus increase the renal capacity for Pi excretion, an effect which is not mediated through cAMP production or sodium handling.

Therefore physiological changes in the endogenous production of 1,25-$(OH)_2D$ have important effects on various parameters of calcium and phosphate metabolism.

References

1 *DeLuca, H.F. and Schnoes, H.K.:* Metabolism and mechanism of action of vitamin D. A. Rev. Biochem. *45:* 631–666 (1976).

2 *Norman, A.W. and Henry, H.L.:* 1,25-Dihydroxyvitamin D_3, a hormonally active form of vitamin D_3. Recent Prog. Horm. Res. *30:* 431–480 (1974).

3 *Kodicek, E.:* The story of vitamin D from vitamin to hormone. Lancet *i:* 325–329 (1974).

4 *Bonjour, J.-P.; Rizzoli, R.; Frey, P.; Haldimann, B.; Hugi, K., and Fleisch, H.:* Influence of chemically synthetized vitamin D_3 metabolites on calcium metabolism in rats. Calcif. Tiss. Res. *22:* suppl., 446–448 (1977).

5 *Morgan, D.B.; Gasser, A.; Largiadèr, U.; Jung, A., and Fleisch, H.:* Effects of a diphosphonate on calcium metabolism in calcium-deprived rats. Am. J. Physiol. *228:* 1750–1756 (1975).

6 *Haldimann, B.; Bonjour, J.-P., and Fleisch, H.:* Role of parathyroid hormone in regulation of main calcium fluxes in rats. Am. J. Physiol. *232:* E535–E541 (1977).

7 *Sammon, P.J.; Staceay, R.E., and Bronner, F.:* Role of parathyroid hormone in calcium homeostasis and metabolism. Am. J. Physiol. *218:* 479–485 (1970).

8 *Aubert, J.-P. et Milhaud, G.:* Méthodes de mesure des principales voies du métabolisme calcique chez l'homme. Biochim. biophys. Acta *39:* 122–139 (1960).

9 *Bonjour, J.-P.; Trechsel, U.; Fleisch, H.; Schenk, R.; DeLuca, H.F., and Baxter, L.A.:* Action of 1,25-dihydroxyvitamin D_3 and a diphosphonate on calcium metabolism in rats. Am. J. Physiol. *229:* 402–408 (1975).

10 *Raisz, L.G.; Trummel, C.L.; Holick, M.F., and DeLuca, H.F.:* 1,25-Dihydroxycholecalciferol: a potent stimulator for bone resorption in tissue culture. Science *175:* 768–769 (1972).

11 *Reynolds, J.J.; Holick, M.F., and DeLuca, H.F.:* The role of vitamin D metabolites in bone resorption. Calcif. Tiss. Res. *12:* 295–301 (1973).

12 *Rizzoli, R.; Fleisch, H., and Bonjour, J.-P.:* Effect of thyroparathyroidectomy on calcium metabolism in rats: role of 1,25-dihydroxyvitamin D_3. Am. J. Physiol. *233:* E160–E164 (1977).

13 *Garabedian, M.; Holick, M.F.; DeLuca, H.F., and Boyle, I.T.:* Control of 25-hydroxycholecalciferol metabolism by parathyroid glands. Proc. natn. Acad. Sci. USA *69:* 1673–1676 (1972).

14 *Fraser, D.R. and Kodicek, E.:* Regulation of 25-hydroxycholecalciferol-1-hydroxylase activity in kidney by parathyroid hormone. Nature new Biol. *241:* 163–166 (1973).

15 *Garabedian, M.; Tanaka, Y.; Holick, M.F., and DeLuca, H.F.:* Response of intestinal calcium transport and bone calcium mobilization to 1,25-dihydroxyvitamin D_3 in thyroparathyroidectomized rats. Endocrinology *94:* 1022–1027 (1974).

16 *Bonjour, J.-P.; Fleisch, H., and Trechsel, U.:* Calcium absorption in diphosphonate-treated rats. Effect of parathyroid function, dietary calcium and phosphorus. J. Physiol., Lond. *264:* 125–139 (1977).

17 *Neer, R.M.; Holick, M.F.; DeLuca, H.F., and Potts, J.T., jr.:* Effect of 1α-hydroxyvitamin D_3 and 1,25-dihydroxy-vitamin D_3 on calcium and phosphorus metabolism in hypoparathyroidism. Metabolism *24:* 1403–1413 (1975).

18 *Russell, R.G.G.; Walton, R.J.; Smith, R.; Preston, C.; Basson, R.; Henderson, R.G., and Norman, A.W.:* 1,25-Dihydroxycholecalciferol and 1α-hydroxycholecalciferol in hypoparathyroidism. Lancet *ii:* 14–17 (1974).

19 *Hugi, K.; Preston, C.; Fleisch, H., and Bonjour, J.-P.:* Renal handling of calcium in rats. Influence of parathyroid hormone and 1,25-dihydroxyvitamin D_3; in Vitamin D. Biochemical, chemical and clinical aspects related to calcium metabolism, pp. 433–435 (De Gruyter, New York 1977).

20 *Rizzoli, R.; Fleisch, H., and Bonjour, J.-P.:* Role of 1,25-dihydroxyvitamin D_3 on intestinal phosphate absorption in rats with a normal vitamin D supply. J. clin. Invest. *60:* 639–647 (1977).

21 *Chen, T.C.; Castillo, L.; Korycka-Dahl, M., and DeLuca, H.F.:* Role of vitamin D metabolites in phosphate transport of rat intestine. J. Nutr. *104:* 1056–1060 (1974).

22 *Walling, M.W. and Kimberg, D.V.:* Effects of 1α,25-dihydroxyvitamin D_3 and *Solanum glaucophyllum* on intestinal calcium and phosphate transport and on plasma Ca, Mg and P levels in the rat. Endocrinology *97:* 1567–1576 (1975).

23 *Walling, M.R.; Kimberg, D.V.; Wasserman, R.H., and Feinberg, R.R.:* Duodenal active transport of calcium and phosphate in vitamin D-deficient rats. Effects of nephrectomy, *Cestrum diurnum* and 1α,25-dihydroxyvitamin D_3. Endocrinology *98:* 1130–1134 (1976).

24 *Tanaka, Y.; Frank, H., and DeLuca, H.F.:* Intestinal calcium transport. Stimulation by low phosphorus diets. Science *181:* 564–566 (1973).

25 *Hughes, M.R.; Brumbaugh, P.F., and Haussler, M.R.:* Regulation of serum 1α,25-

dihydroxyvitamin D_3 by calcium and phosphate in the rat. Science *190:* 578–579 (1975).
26 *Hill, L.F.; Lumb, G.A.; Mawer, E.B., and Stanbury, S.W.:* Indirect inhibition of the synthesis of 1,25-dihydroxycholecalciferol in rats treated with a diphosphonate. Clin. Sci. *44:* 335–347 (1973).
27 *Baxter, L.A.; DeLuca, H.F.; Bonjour, J.-P., and Fleisch, H.:* Inhibition of vitamin D metabolism by ethane-1-hydroxy-1,1-diphosphonate. Archs Biochem. Biophys. *164:* 665–672 (1974).
28 *Bonjour, J.-P.; Preston, C., and Fleisch, H.:* Effect of 1,25-dihydroxyvitamin D_3 on the renal handling of Pi in thyroparathyroidectomized rats. J. clin. Invest. *60:* 1418–1428 (1977).
29 *Bonjour, J.-P. and Fleisch, H.:* The effect of vitamin D and its metabolites on the renal handling of phosphate; in Vitamin D. Biochemical, chemical and clinical aspects related to calcium metabolism, pp. 419–431 (De Gruyter, New York 1977).
30 *Puschett, J.B.; Fernandez, P.C.; Boyle, I.T.; Gray, R.W.; Omdahl, J.L., and DeLuca, H.F.:* The acute renal tubular effects of 1,25-dihydroxycholecalciferol. Proc. Soc. exp. Biol. Med. *141:* 379–384 (1972).
31 *Popovtzer, M.M.; Robinette, J.B.; DeLuca, H.F., and Holick, M.F.:* The acute effect of 25-hydroxycholecalciferol on renal handling of phosphorus. Evidence for a parathyroid hormone-dependent mechanism. J. clin. Invest. *53:* 913–921 (1974).

Dr. *H. Fleisch,* Department of Pathophysiology, University of Berne, Murtenstrasse 35, *CH–3010 Berne* (Switzerland)

Contr. Nephrol., vol. 13, pp. 104–114 (Karger, Basel 1978)

Endocrinological Aspects of PTH Metabolism in the Kidney[1]

R.-D. Hesch, H. Ebel, R. Hehrmann and H. Jüppner

Department of Endocrinology, Medizinische Hochschule, Hannover, and Department of Physiology, Freie Universität, Berlin

The kidney has a major role in the calcium metabolism, most of which is mediated by hormonal actions. It plays, however, a double role: (1) kidney is a principal target organ whereby the interaction of parathyroid hormone (PTH) and 25-OHD_3 is the most effective site of action, and (2) kidney takes part in the metabolism of both hormones; PTH is transformed to different peptides and 25-OHD_3 to metabolites like 1,25-OH_2D_3 and others.

The aim of our considerations is to demonstrate that peripheral handling at particular receptors, the occurrence of either biological active or inactive and/or regulating PTH metabolites, and immunological processes are important to understand bioexpression of PTH. One of the main interests in PTH action is its binding to specific receptors, as recently demonstrated by *DiBella et al.* (6), *Malbon and Zull* (16) and *Heath and Aurbach* (10). These studies deal probably with large nonspecific binding of biological inert PTH. PTH exerts important biological effects along the nephron and undergoes furthermore dramatic metabolic transformations which are tentatively summarized in figure 1.

It is not yet possible to directly relate binding, metabolism and functional processes to each other. We will thus consider (1) binding of PTH to distinct structures of the nephron, (2) PTH transformation at the receptor site, and (3) immunological reactions of the receptor. We were recently able to develop a new assay technique to observe receptor-bound hormone. It avoids any labelling of the hormones and thus preserves the biological conformation of the respective hormone, hence allowing to observe PTH at the physiological range on its receptor (12–14, 17, 18).

[1] Supported by the Deutsche Forschungsgemeinschaft He 593/9.

Glomerular Binding of PTH

Low molecular weight plasma particles (<50,000 m.w.) readily traverse glomerular filters and are therefore taken up and catabolized in the tubular cells as recently reviewed by *Strober and Waldmann* (21). *Ardaillou* and coworkers from Paris, in cooperation with our group, were recently able to show that tritiated biologically active 1-34 PTH exhibiting binding to isolated glomeruli displaces intact 1-84 PTH bound to the same target receptor. The physiological significance of this receptor in view of glomerular filtration of PTH is uncertain.

Assuming however a perfusion rate of 600 ml/min for both kidneys and a concentration of about 10 pg/ml intact PTH, the data indicate that glomeruli are exposed to about 8.5 μg of intact PTH/24 h (19). Assuming further a secretion rate of 50 μg/24 h of intact PTH, roughly 20% of whole intact PTH are subject to glomerular handling. This means that catabolism of PTH cannot entirely be located at this site of hormonal action.

Tubular Receptors

More data are available on receptors in cortical plasma membranes, the stimulation of adenylate cyclase and consequent protein kinase activation. There are, however, many unanswered questions, which concern mainly the distribution of adenylate cyclase, of cAMP-dependent protein kinase (4, 6, 20) and also transport systems like reabsorption of phosphate, calcium and potassium but

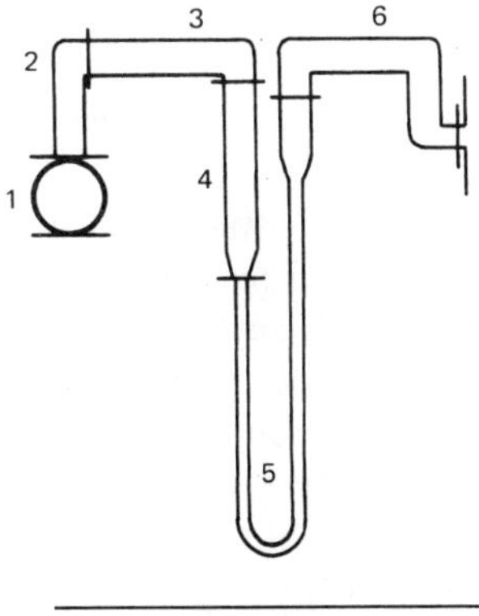

Functions of PTH along the nephron

1 Receptor like binding, stimulation of adenylate cyclase
filtration of intact PTH and fragments

2 Inhibition of P_i reabsorption > in 5
Inhibition of fluid absorption
Inhibition of sodium and calcium absorption

2,3,6, (4) Receptor like binding onto basal-lateral membranes and brush-borders
Stimulation of adenylate cyclase
Inactivation and fragmentation of intact PTH
Variation in enzyme stimulation along the length of the tubule
Stimulation of 1-α- hydroxylase of 25-OHD_3
Stimulation of gluconeogenesis

5 Inhibition of P_i reabsorption < 2

5, (6) Enhancement of calcium reabsorption
(less than in 2)

Fig. 1. Schematic representation of some biological functions mediated by PTH along the nephron. For details see text and references.

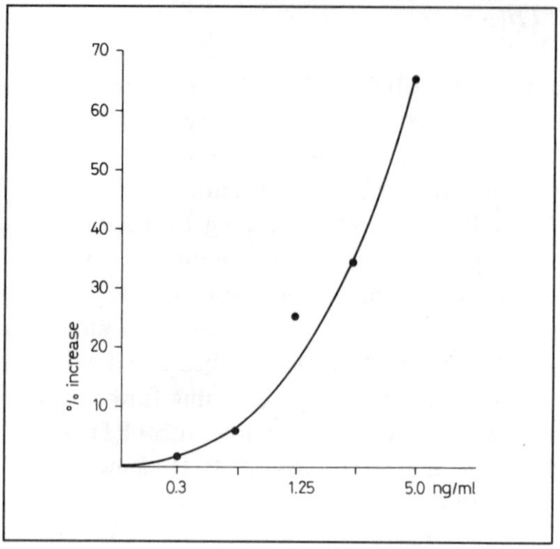

Fig. 2. Binding of PTH (bPTH 71/324 MRC London) onto brush border membranes. The assay conditions are as described for basal membranes (18). Note that even 500 pg/ml exhibit significant binding. Saturation is not reached with 5.0 ng/ml which may be due to high enzymatic PTH turnover or large low affinity binding sites. Although binding is specific for PTH due to the intrinsic specificity of the assay design, this does not mean that the binding is followed by a hormonal stimulatory function known as specific for PTH.

also the locus of PTH-dependent tubular gluconeogenesis (1, 2, 9). The heterogeneous pattern of distribution of these biochemical processes along the tubules and the different time of maximal reaction between occurrence of cAMP stimulation and dependent or independent other activations by PTH suggest rather separate PTH-mediated effects not only at different sites along the nephron but also at the cellular sites itself. We have now accumulated increasing evidence of specific binding of 1-84 PTH onto rat kidney brush border membranes. Moreover PTH is metabolized by brush borders (fig. 2).

It is evident that PTH peptides bound and metabolized by a luminal receptor may enter the cell by pinocytosis. This may help to explain observations of *Zanelli* (22) on heterogeneous stimulation of enzymes intracellularly along the nephron by cytochemical methods. Specific brush border binding and uptake of PTH peptides explains also results of *Shlatz et al.* (20) and *Liang and Sacktor* (15) as well, because adenylate cyclase which is probably not present in brush borders may not be a prerequisite for possible intracellular effects of PTH peptides taken up from the luminal receptor site. Most of our knowledge stems from studies of antiluminal binding of PTH on basal lateral membranes. Intact PTH is bound to those membranes with high affinity.

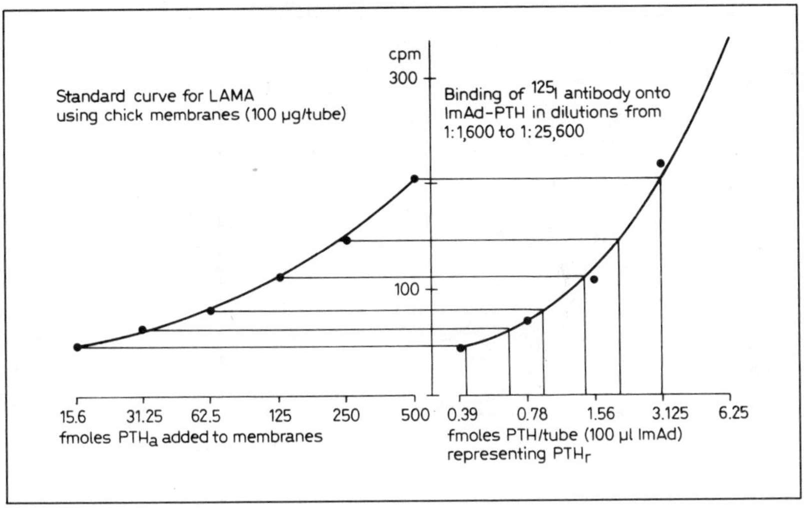

Fig. 3. Left: If increasing concentrations of ambient PTH_a are incubated with chick membranes, absolute receptor bound hormone is observed by the increase of labelled antibody bound to the receptor bound PTH_r. This increase is compared to the binding of identical antibody in a simultaneous assay to a delution curve of solid phase PTH.

To calculate this important parameter we determined absolute receptor-bound hormone by a new direct approach. We found that at ambient concentration of PTH in the femtomole range receptor bound hormone amounts to attomoles (fig. 3). Since PTH is metabolized at the receptor it was necessary to correct for this process which was omitted in all former studies. The three major issues addressed in these studies include the following (fig. 4).

Without correction for inactivation of bioactive PTH, the Scatchard plot would suggest receptor heterogeneity, positive cooperativity of a 2-step reaction, as erroneously described for many receptors (3). If correction is made for intact PTH retaining immunochemically required conformation, less but still evident deviation from the ideal curve is obtained. Only after correction for conformationally and biologically intact PTH detectable in the LAMA system, a classical kinetic is reached with an equilibrium dissociation constant of 1.1×10^{-13} *M* and a binding capacity of 5.8×10^{-15} *M*. With regard to these issues several conclusions are yet possible: (1) PTH binds with high affinity and limited capacity onto tubular membranes; (2) this reaction obeys a single classical kinetic if appropriate correction for hormone inactivation is made; (3) the high affinity receptor and the PTH metabolizing enzyme are not separate entities, and (4) the stimulation of adenylate cyclase depends upon the functional state of this PTH-receptor complex.

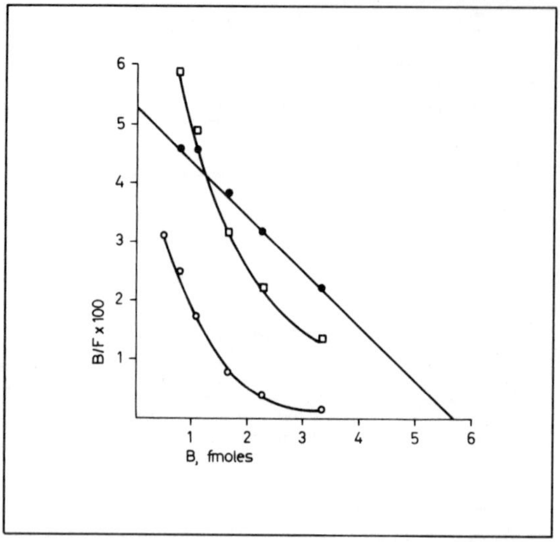

Fig. 4. Scatchard plot for the kinetic of receptor-exposed and absolute receptor-bound hormone. ○ = Representation of data without correction for receptor-exposed hormone. □ = Correction for hormone inactivation observing alterations of the immunochemical conformation of 1-84 PTH. ● = A straight and classic kinetic of first order is only obtained when correction is made for changes of hormonal conformation required to bind onto receptor membranes.

It is evident that the rate of association and dissociation of the PTH-receptor complexes depends upon metabolism of bound intact hormone to N-terminal short-lived bioactive and C-terminal bioinactive peptides, because these will regulate ultimately the duration of an appropriate function at the antiluminal site. This has recently been demonstrated in bioassays using different PTH fragments very elegantly by *Parsons et al.* (19), indicating functional heterogeneity similar and possibly linked to that of enzymes.

Goltzman et al. (8), in contrast to other groups, have recently reached the conclusion that cleavage of intact PTH is not required for its bioexpression. Our own results similar to those of *Chu et al.* (5) are somewhat different. Using our most sensitive receptor assay method, we were never able to simulate assay conditions under which intact PTH would react with its receptor without being metabolized. To describe the pattern of PTH peptides emerging from reaction with basal-lateral membranes we have followed it prior to and after chromatography on Biogel P_{10}, whereby PTH was determined by region-specific assays developed in our laboratories. It was found that receptor-detectable PTH was rapidly inactivated, followed by 1-84 immunochemically intact PTH similar to 1-34 PTH, whereas C-terminal PTH was only minimally altered (fig. 5).

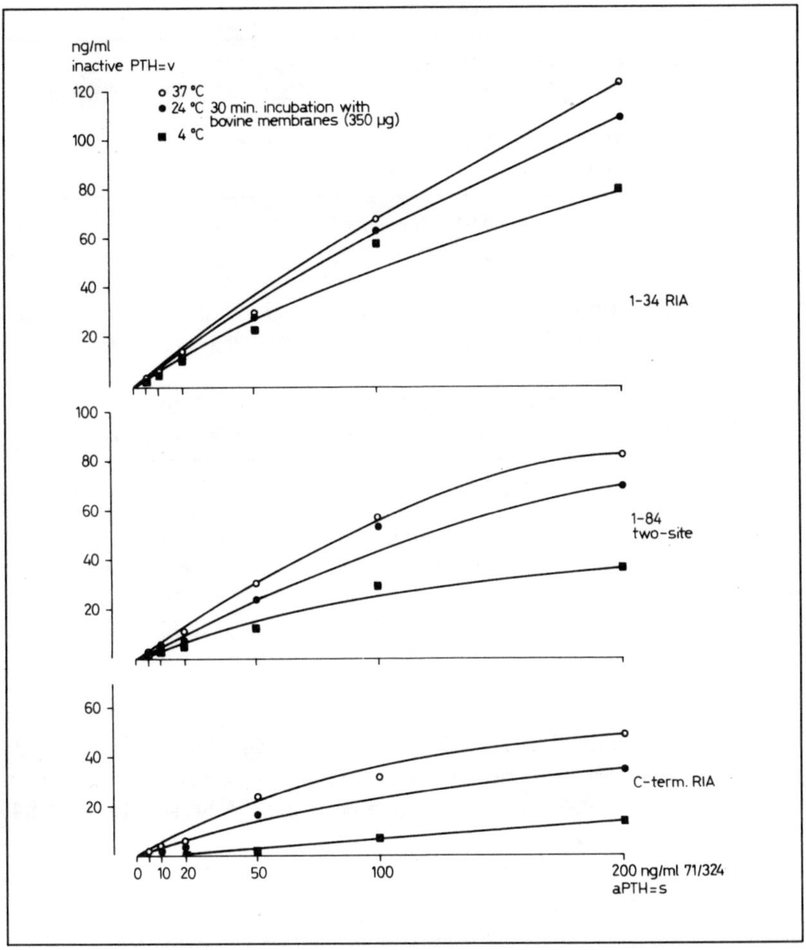

Fig. 5. Inactivation of 1-84, 1-34 and C-terminal PTH at different temperatures by bovine renal plasma membranes. For 1-34 peptide no saturation is obtained but even for 1-84 and C-terminal PTH saturation is not complete. V_{max} and K_m could not be determined.

Direct exposure of 1-34 PTH to membranes revealed that only 50% were destroyed after 1 h, indicating much higher stability of the 1-34 peptide than of the 1-84 PTH. Kinetic studies are required to further understand these observations. When data from chromatography are listed (table I) in groups of PTH peptides with different molecular weight, the following can be said:

(1) In the high molecular weight fraction 1-84 and C-terminal immunoreactive PTH is cleaved to 48 and 56%, respectively, and N-terminal 1-34

Table I

	MW 13,000–9,000 fractions 29–35			MW 9,000–6,500 fractions 36–41			MW 6,500–4,500 fractions 42–46			MW 4,500–1,500 fractions 47–63		
	1-84	C-term	1-34	1-84	C-term	1-34	1-84	C-term	1-34	1-84	C-term	1-34
Native bPTH ng = 100%	17.4	271.5	–	8.97	266.0	13.7	3.50	117.0	25.1	1.73	39.0	101.4
Bovine memb.	9.65	171.6	19.7	5.17	224.5	14.3	1.50	114.7	39.3	0.53	48.5	75.3
% of native	55%	63%	–	58%	84%	104%	43%	98%	157%	31%	124%	74%
Rat BLM	8.4	137.5	5.3	4.1	147.0	19.5	1.1	66.5	29.4	0.45	46.5	87.4
% of native	48%	56%	–	46%	55%	142%	31%	57%	117%	26%	119%	86%

5 μg of bovine PTH were incubated for 30 min at 37°C with 350 μg bovine renal plasma membranes and 50 μg of rat basal lateral membranes (BLM) followed by subsequent chromatography on Biogel P_{10}. The elution pattern is listed in groups of different molecular weight expressing the resulting data from the determination of 1-84, 1-34 and C-terminal PTH after integration. Absolute concentrations and the percentage of the original PTH prior to chromatography are given.

fragments are generated. (2) A 50% decrease of 1-84 PTH is seen in the 9,000–6,500 fractions with the appearance of N-terminal 1-34 PTH peptides. (3) In the 6,500 to 4,500 molecular weight fractions higher 1-34 and lower 1-84 and C-terminal PTH concentrations are present before incubation by which N-terminal fragments are additionally generated. (4) In the 4,500–1,500 molecular weight fractions the highest concentration of 1-34 PTH is seen, but which is decreased by incubation, while small C-terminal fragments increase.

Therefore it is obvious that basal-lateral membranes produce a variety of N-terminal and C-terminal peptides. Since they bind for different time periods to the receptor due to their individual metabolism, occupancy and enzyme activity of the receptor are determined by these substrates. Recent studies in our laboratory on influences of PTH peptides upon kinetics of PTH on renal membranes and intact lymphocytes revealed different effects on binding sites and the enzymatic subunit in the membrane (11). We conclude that a complex interaction of N- and C-regional peptides, coming from the luminal and antiluminal site of specific tubular receptors, will ultimately determine bioexpression of PTH along the nephron. Luminal PTH does possibly not need membrane-bound cyclase systems, in contrast to basal-lateral membranes whose PTH effects are mediated by stimulation of the enzyme (fig. 6).

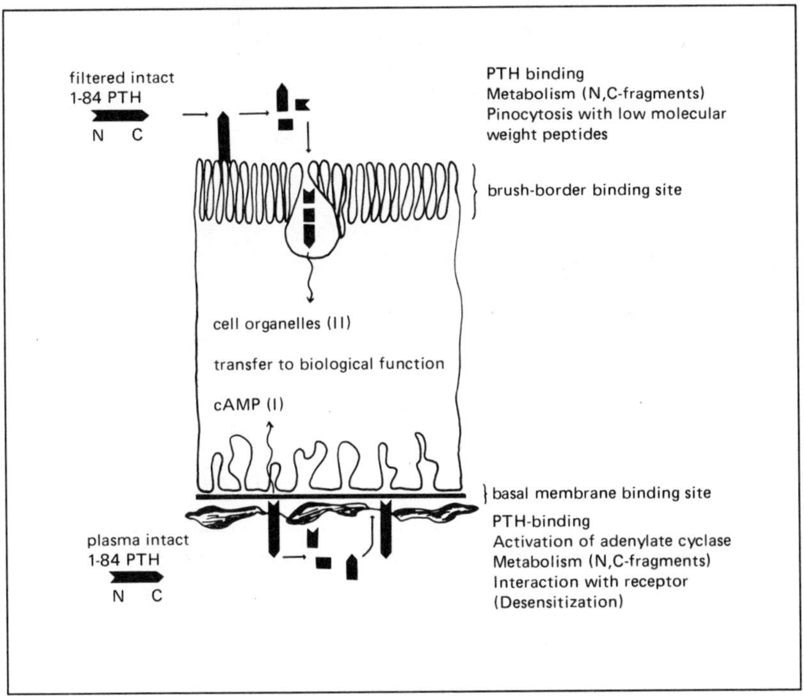

Fig. 6. Model of our current understanding of the two functional sites of PTH. For details of the brush border action of PTH and that of antiluminal basal membrane see text.

Finally we would like to propose a further entity of immune disease. We know about the discrepancies interpreting radioimmunochemical PTH determinations in uremia. They seem to be poorly correlated to the clinical state of hyperparathyroidism, for example to quantitative results of bone biopsies. We were recently able to demonstrate that antibodies raised against basal-lateral membranes and against brush borders inhibit binding of PTH to partially purified tubular membranes specifically (fig. 7). This situation is very similar to that observed in insulin-resistant diabetes mellitus, myasthenia gravis and thyrotoxicosis. In 10 uremic patients with inappropriate high PTH values we were able to detect a serum factor inhibiting binding of PTH to its receptor. This factor has been shown to be a globulin. Further studies will indicate the mechanisms by which resistance to PTH is due to the yet fourth example of a 'hormone receptor antibody disease' (fig. 8).

In conclusion PTH handling at its different receptor sites opens a new field of understanding the numerous bioeffects of this hormone in renal physiology and diseases.

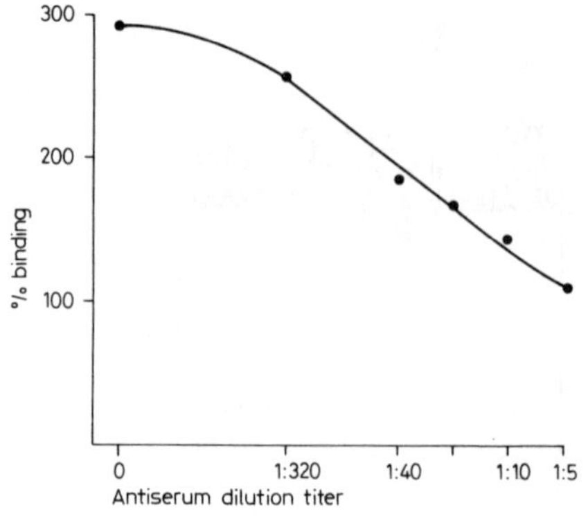

Fig. 7. Antiserum against basal-lateral membrane is raised in rabbits after one multiple site immunization of solubilized (0.5% Triton-X_{100}) membranes mixed with complete Freund adjuvant (v/v). Binding of 50 ng PTH onto membranes is inhibited by serial dilutions of the antiserum.

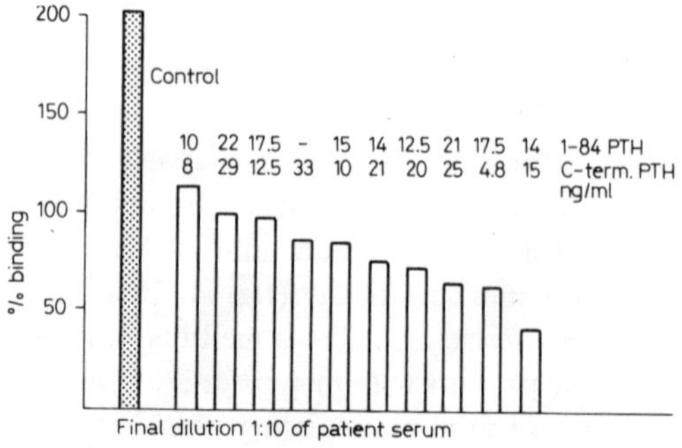

Fig. 8. In 10 uremic patients inhibition of PTH onto bovine membranes is demonstrated. Plasma is diluted 1:10 with assay buffer. No correlation to endogenous 1-84 or C-terminal plasma levels was found. The patient with the greatest inhibition did not exhibit histological signs of hyperparathyroidism.

Summary

Various methods for the measurement of intact PTH, PTH fragments and PTH binding to receptors were applied to clarify the role of renal receptors for the bioexpression of PTH. Glomerular receptors may contribute approximately 20% of the renal PTH catabolism. PTH also binds to tubular receptors at the luminal side (brush border membranes) as well as at the antiluminal site (basal-lateral membrane).

Scatchard plot analysis of PTH binding to these receptors allows the calculation of an equilibrium dissociation constant and binding capacity, if correction is made for inactivation of bioactive PTH (measured by LAMA). Binding to receptors is – in our systems – invariably associated with degradation of the intact hormone to N-terminal and C-terminal fragments of different molecular weight. Antibodies against basal lateral membranes and against brush border membranes are able to inhibit PTH binding to tubular membranes. Sera of uremic patients with inappropriately high PTH contain a globulin which interferes with PTH binding to the receptor, suggesting that uremia may be another example for a 'hormone receptor antibody disease'.

References

1 *Agus, Z.S.; Gardner, L.B.; Beck, L.H., and Goldberg, M.:* Effects of parathyroid hormone on renal tubular reabsorption of calcium, sodium and phosphate. Am. J. Physiol. *224:* 1143–1148 (1973).

2 *Brunette, M.G.; Taleb, L., and Carriere, S.:* Effect of parathyroid hormone on phosphate reabsorption along the nephron of the rat. Am. J. Physiol. *225:* 1076–1081 (1973).

3 *Boeynaems, J.M. and Dumont, J.E.:* Quantitative analysis of the binding of ligands to their receptors. J. cycl. Nucl. Res. *1:* 123–142 (1975).

4 *Chabardès, D.; Imbert, M.; Clique, A.; Montégut, M., and Morel, F.:* PTH sensitive adenylate cyclase activity in different segments of the rabbit nephron. Pflügers Arch. *354:* 229–239 (1975).

5 *Chu, L.L.H.; Forte, L.; Anast, C.S., and Cohn, D.V.:* Interaction of parathyroid hormone with membranes of kidney cortex: degradation of the hormone and activation of adenylate cyclase. Endocrinology *97:* 1014–1013 (1975).

6 *Forte, L.R.; Chao, W.H.; Walkenbach, R.J., and Byington, K.H.:* Studies of the kidney plasma membrane adenosine-3',5'-monophosphate dependent protein kinase. Biochim. biophys. Acta *385:* 84–96 (1975).

7 *DiBella, J.P.; Dousa, T.P.; Miller, S.S., and Arnaud, C.D.:* Parathyroid hormone receptors of renal cortex: specific binding of biologically active, ^{125}I-labelled hormone and relationship to adenylate cyclase activation. Proc. natn. Acad. Sci. USA *71:* 723–726 (1974).

8 *Goltzman, D.; Peytremann, A.; Callahan, E.N.; Segre, G.V., and Potts, J.:* Metabolism and biological activity of parathyroid hormone in renal cortical membranes. J. clin. Invest. *57:* 8–19 (1976).

9 *Guder, W.:* Hormonal regulation of renal gluconeogenesis in isolated tubule fragments; in Current problems in clinical biochemistry, vol. 6, pp. 202–213 (Huber, Bern 1975).

10 *Heath, D.A. and Aurbach, G.D.:* Studies on the binding of ^{125}I-parathyroid hormone to renal cortical membranes. Excerpta Med. Int. Congr. Series No. 346, pp. 159–162 (Excerpta Medica, Amsterdam 1974).

11 *Hesch, R.-D.; Biallasiewic, A., and Jüppner, H.:* Unpublished results.
12 *Jüppner, H. and Hesch, R.-D.:* Inhibition of PTH receptor binding and PTH mediated adenylate cyclase activity by somatostatin. Biochem. biophys. Res. Commun. *72:* 945–951 (1976).
13 *Jüppner, H. and Hesch, R.-D.:* Enzymatic modification of bovine and ovine parathyroid receptor. FEBS Lett. *67:* 281–283 (1976).
14 *Jüppner, H.; Ebel, H., and Thiele, J.:* Characterization of the PTH inactivating enzyme. Acta endocr., Copenh. *84:* suppl. 208, pp. 100–101 (1977).
15 *Liang, C.T. and Sacktor, B.:* Preparation of renal cortex basal-lateral and brush border membranes. Localization of adenylate cyclase and guanylate cyclase activities. Biochim. biophys. Acta *426:* 474–487 (1977).
16 *Malbon, C.C. and Zull, J.E.:* Interactions of parathyroid hormone and plasma membranes from rat kidney. Biochem. biophys. Res. Commun. *56:* 952–958 (1974).
17 *McIntosh, C.H.S. and Hesch, R.-D.:* Labelled antibody membrane assay for parathyroid hormone. A new approach to the measurement of receptor bound hormone. Biochem. biophys. Res. Commun. *64:* 376–383 (1975).
18 *McIntosh, C.H.S. and Hesch, R.-D.:* Characterization of the parathyroid receptor in renal plasma membranes by labelled hormone and labelled antibody binding techniques. Biochim. biophys. Acta *426:* 535–546 (1976).
19 *Parsons, J.A.; Rafferty, B.; Gray, D.; Reit, B.; Zanelli, J.M.; Keutmann, H.T.; Tregear, G.W.; Callahan, E.N., and Potts, J.T.:* Pharmacology of parathyroid hormone and some of its fragments and analogues. Excerpta Med. Int. Congr. Series No. 346, pp. 33–39 (Excerpta Medica, Amsterdam 1975).
20 *Shlatz, L.J.; Schwartz, I.; Kinne-Saffran, E.M., and Kinne, R.:* Distribution of parathyroid hormone-stimulated adenylate cylcase in plasma membranes of cells of the Kidney cortex. J. Membr. Biol. *24:* 131–144 (1975).
21 *Strober, W. and Waldmann, T.A.:* The role of the kidney in the metabolism of plasma proteins. Proc. 6th Int. Congr. Nephrol., Florence, 1975, pp. 392–405 (Karger, Basel 1976).
22 *Zanelli, J.:* Ultrasensitive cytochemical localization of parathyroid hormone action on the kidney. 6th Parathyroid Conf., Vancouver 1977 (in press).

Priv.-Doz. Dr. *R.-D. Hesch,* Departement für Innere Medizin, Medizinische Hochschule Hannover, Karl-Wiechert-Allee 9, *D–3000 Hannover 61* (FRG)

Contr. Nephrol., vol. 13, pp. 115–124 (Karger, Basel 1978)

Kinetics of PTH Metabolism and PTH Fragments in Chronic Renal Failure

N. Lustenberger, R. Hehrmann, H. Jüppner, G. Delling and R.D. Hesch
Department Innere Medizin, Medizinische Hochschule, Hannover, und Pathologisches Institut der Universitätsklinik Eppendorf, Hamburg

Introduction

It is generally assumed that parathyroid hormone (PTH) secreted by the glands is predominantly the monochain of intact 1-84 PTH (1, 20). Recent studies on subcellular fractions of human parathyroid adenomata by gel filtration techniques and radioimmunoassays using antibodies against carboxyl (C)-regional and amino (N)-regional parts of the molecule have, however, demonstrated peptide fragments within the gland, indicating glandular heterogeneity of PTH. On the other hand, plasma PTH peptide families result not only from secretion by the glands, but also from metabolic turnover of glandular PTH peptides (fig. 1). Plasma levels of the intact 1-84 PTH are extremely low due to rapid binding to receptors in its main target organs: bone, kidney and liver (5, 16, 24) and cleavage to its peptide degradation products. This mainly takes place in the liver (3, 13) and in the kidney (4, 8, 10–13, 17–19). Binding of the intact 1-84 PTH to these organs implicits degradation as recently demonstrated by our group (10). Because of extremely low concentrations in the plasma, N-regional PTH peptides are difficult to detect, while C-regional peptides may partly diffuse back from receptor into the plasma and can easily be assayed.

For low-molecular weight proteins, the kidney is regarded to be a primary organ of catabolism by means of glomerular filtration and tubular metabolism as discussed by *Strober and Waldmann* (23). In chronic renal failure increased plasma levels of low-molecular weight proteins ('middle molecules', MW 350–2,000) (2) are assumed due to reduced glomerular filtration rate (GFR). So the question is, if impaired metabolism of PTH peptide families contributes to the high levels found in almost any uremic patient, sometimes lacking clinical evidence for the presence of biologically active secondary hyperparathyroidism.

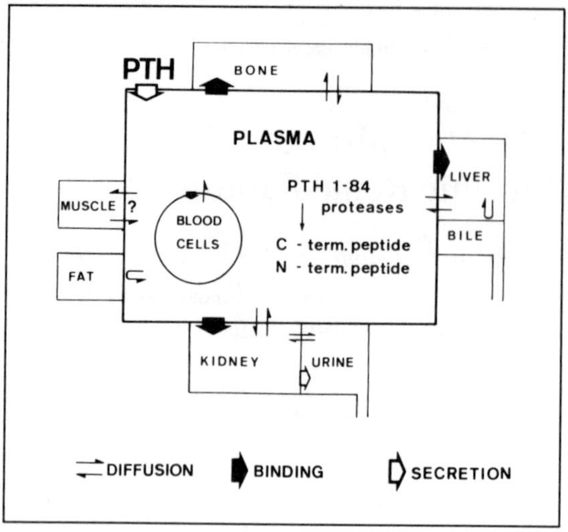

Fig. 1. Regulation of plasma PTH levels.

Controversial results about the influence of impaired renal function on PTH metabolism are obtained from studies in man after parathyroidectomy (14, 15) and in dogs (9, 22). The present study was undertaken to evaluate the role of the kidney in PTH turnover in man using specific assays for intact 1-84 PTH, C-regional and N-regional fragments.

Material and Methods

This study was performed in randomly selected patients suffering from chronic renal diseases GFR 15–30 ml/min, n = 4; GFR ⩽ 10 ml/min, n = 36, including 3 anephric, bilateral nephrectomized patients, the others on a predialysis treatment regimen. No selection was done for signs or symptoms of overt renal osteopathy. Phosphate-binding drugs and oral calcium supplements were given if needed, none had vitamin D treatment.

The following parameters were studied: *Serum:* (total) calcium, phosphate, alkaline phosphatase, immunoreactive PTH (intact 1-84 PTH; C-regional (C-RIA) and N-regional 1-34 peptide fragment). *Iliac crest bone biopsy:* quantitative histomorphometry for signs of osteomalacia and hyperparathyroidism (HPT). *Kinetics of bovine PTH (bPTH) distribution and metabolism:* 400 U of bPTH (from Hormon-Chemie, München) were infused with a Braun constant infusion pump over 20 min and blood samples drawn at short time intervals up to 120 min. PTH assays were performed as described by us previously (6, 7).

Polyfructosan (Inutest®, Deutsche Laevosan-Gesellschaft), having about the same molecular weight (MW 5,200) as PTH peptides, was infused simultaneously. Thus the period of distribution in the extracellular space can be separated from predominant glomerular filtration and metabolic turnover, respectively. For control, kinetics of bPTH distribution and metabolism were performed in 12 healthy volunteers.

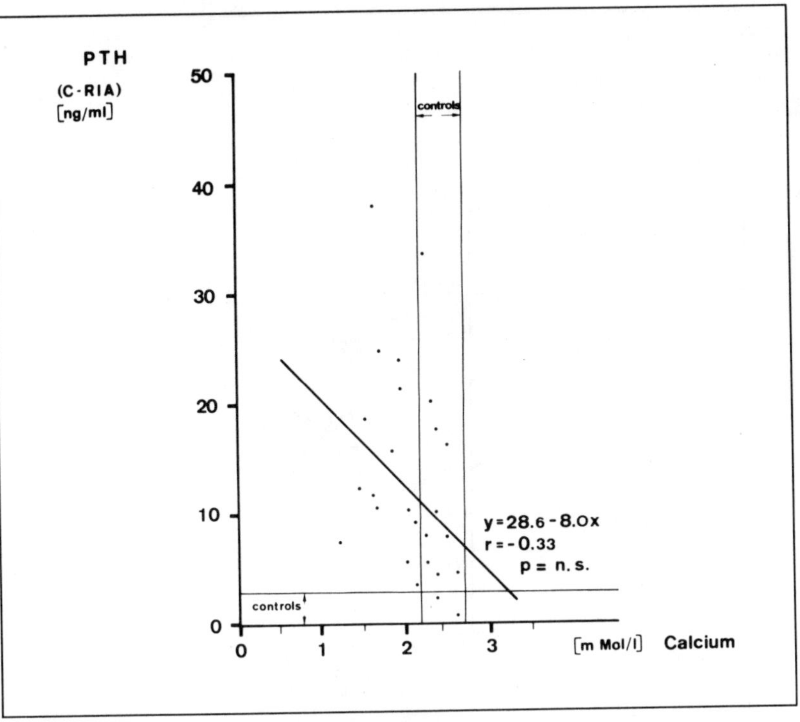

Fig. 2. Correlation between plasma PTH (C-regional fragment) and total serum calcium (GFR ⩽10 ml/min).

Results

Serum Parameters and Bone Biopsy

Correlations between serum calcium, alkaline phosphatase, PTH (C-regional peptides) and severity of secondary HPT in iliac crest bone biopsy are summarized in figures 2–4. C-regional PTH showed a moderate negative correlation with total serum calcium (fig. 2), whereas no correlation was found between intact 1–84 PTH or N-regional peptide and calcium. Figures 3 and 4 illustrate the again moderate but significant correlation between alkaline phosphatase and severity of histological signs of HPT (fig. 3) and between C-regional PTH and bone histology (fig. 4).

Kinetic Analyses

When transferred into a semilogarithmic plot, disappearance curves of *inulin* for both controls (fig. 5a) and uremic patients (fig. 5b) exhibit a two-compart-

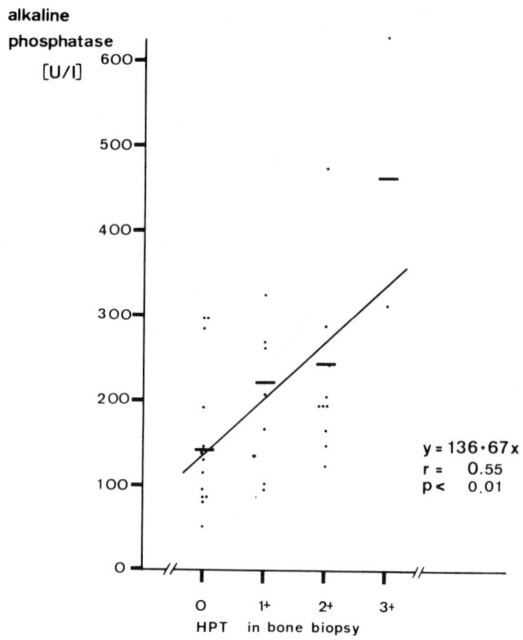

Fig. 3. Correlation between serum alkaline phosphatase and severity of HPT in bone biopsy.

ment system with a first, rapid component which is known to represent distribution and equilibration with interstitial fluid, and a second, slow component representing elimination by glomerular filtration. Extracellular fluid volume showed no significant difference between uremic patients and controls. Half-lives (hl) were calculated from the mean values of all experiments for the first (hl_1: 20–30 min) and second (hl_2: 35–120 min) time period (table I).

Intact 1-84 bPTH. When bPTH infusion is stopped after 20 min, plasma levels of intact 1-84 bPTH show a most rapid decline of all substances under study (fig. 5; table I). In controls, hl_1 is 1.9 min and 1-84 PTH is no more detectable after 30 min. In uremia, disappearance of intact 1-84 bPTH is slightly impaired (hl_1: 5.1 min) and a second component with an apparent hl of 24.4 min becomes detectable.

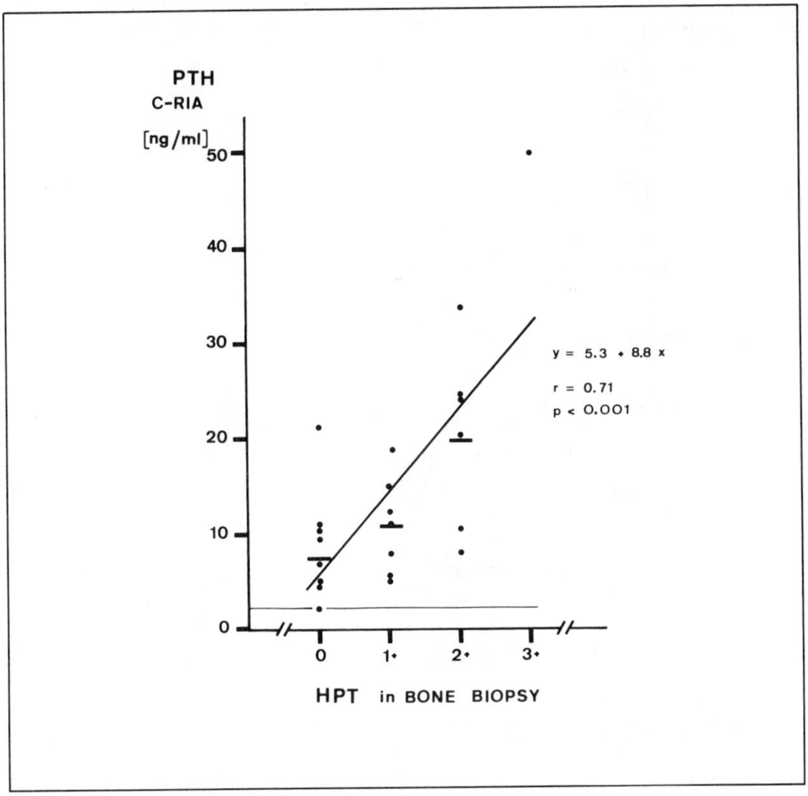

Fig. 4. Correlation between plasma PTH (C-regional fragment) and severity of HPT in bone biopsy.

Carboxyl-regional peptides [*bPTH (C-RIA)*] Plasma levels of this fragment are found about ten times higher than those of intact 1-84 bPTH and N-regional fragment. In uremia, disappearance rate for the first time period ('distribution') is less rapid (hl_1: 11.8 min) and less pronounced than in controls (hl_1: 5.2 min), and cleavage of C-regional peptides is not complete at the end of the experiment. (Controls, $\bar{x} \pm$ SEM: 1.08 ± 0.23 ng/ml; uremia, $\bar{x} \pm$ SEM: 3.10 ± 0.61 ng/ml; $p < 0.005$).

Amino-regional peptides [*bPTH (1-34)*] In controls, N-regional peptides show a second peak in 7 out of 12 experiments, probably by peripheral generation of this fragment in the course of hormone degradation. It remains unexplained whether and why this second peak is absent in some of the controls and most of the uremic patients.

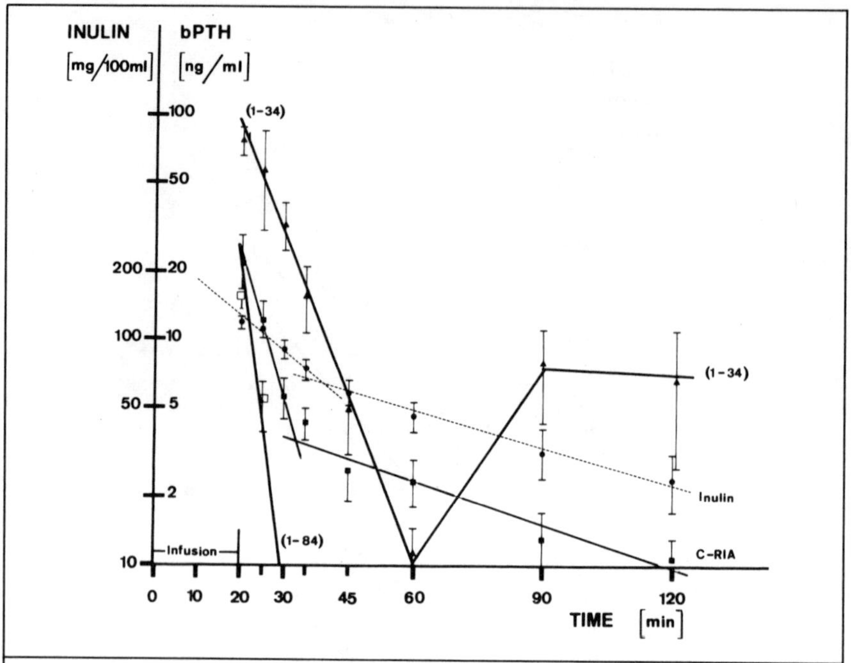
INULIN
[mg/100ml]
bPTH
[ng/ml]
(1-34)
(1-34)
Inulin
C-RIA
(1-84)
Infusion
TIME [min]

5a

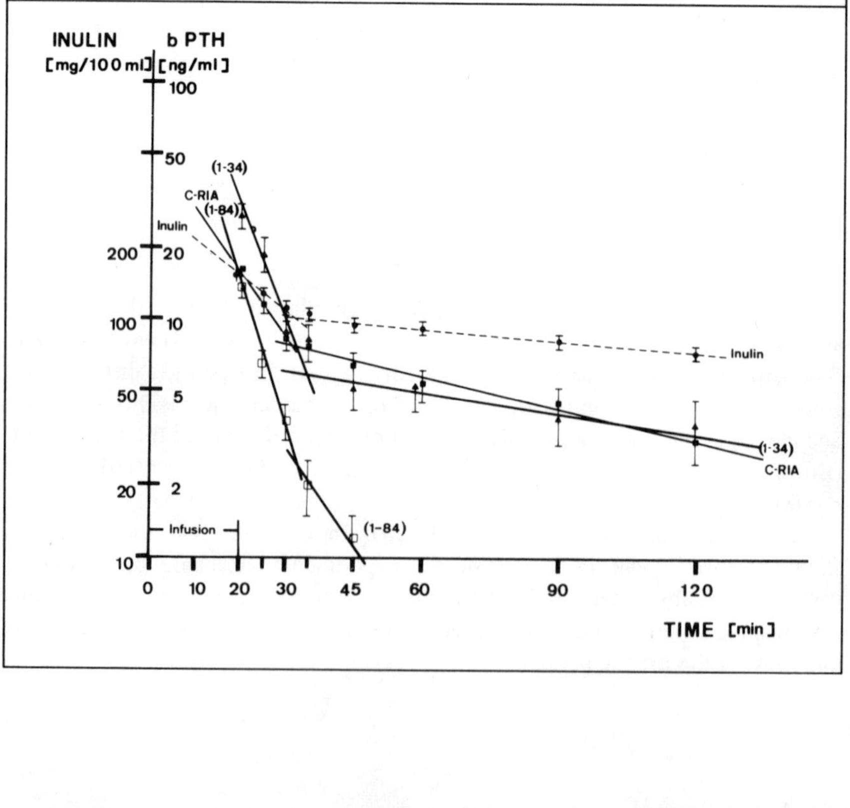
INULIN
[mg/100 ml]
b PTH
[ng/ml]
(1-34)
C-RIA
(1-84)
Inulin
Inulin
(1-34)
C-RIA
(1-84)
Infusion
TIME [min]

5b

Table I. Half-lives of inulin and PTH

	Half-life 1 (20–30 min)				Half-life 2 (35–120 min)			
	inulin	bPTH			inulin	bPTH		
		1–84	C-RIA	1–34		1–84	C-RIA	1–34
Controls (n = 12)	24.7	1.9	5.2	8.0	56.4	n.d.	46.2	2.peak
GFR 15–30 ml/min (n = 4)	31.5	5.5	11.7	6.1	121.6	55.0	50.6	231
GFR ⩽10 ml/min (n = 36)	21.4	5.1	11.8	6.1	173.3	24.4	71.5	86.6
BNX (n = 3)	32.8	2.7	12.2	11.0	182.4	27.2	41.3	36.1

Values calculated from disappearance curves of figure 5. BNX = Bilateral nephrectomized patients; n.d. = not detectable.

Discussion

This study is concerned with observations of endogenous human PTH, calcium, phosphate values in correlation to histomorphometry of the bone. These data are compared to kinetic analyses of infused bPTH.

By testing bPTH and using specific assays for intact hormone, C-regional and N-regional PTH peptides, interference of circulating endogenous PTH is avoided. Previous studies (14, 15) of PTH metabolism in man after partial parathyroidectomy have to take into account peripheral generation of peptide

Fig. 5a. Disappearance curves of infused inulin and bPTH (controls). Semilogarithmic plot versus time, mean values ± SE. ● = Inulin; ■ = bPTH-C-RIA; □ bPTH-(1-84) × 10; ▲ = bPTH-(1-34) × 10.

Fig. 5b. Disappearance curves of infused inulin and bPTH (uremia). Semilogarithmic plot versus time, mean values ± SE. ● = Inulin; ■ = bPTH-C-RIA; □ = bPTH-(1-84) × 10; ▲ = bPTH-(1-34) × 10.

fragments and continuous hormone production by parathyroid gland tissue left *in situ,* thus resulting in prolonged half-lives. Different handling of the alien bovine bPTH in man compared to human PTH is not known, but may be objected.

The data presented here clearly demonstrate a most rapid inactivation of intact 1-84 bPTH in any group of renal function. This suggests that degradation may take place at different sites under different conditions of health and renal disease. A shift from renal to extrarenal target organs must at least be considered as an important possibility in this regard.

Different disappearance curves of PTH fragments in uremia compared to controls must be discussed in more detail: First it is evident that kinetics of intact 1-84 PTH in uremic patients, although slightly impaired, do not exhibit the striking difference seen for N-regional and C-regional PTH peptides. Although supraphysiological doses of bPTH were infused, this means that intact PTH is properly handled in uremic subjects. The second component in uremic subjects may indicate that the amount of endogenous PTH receptors is nevertheless more limited compared to normals. But under physiological conditions a saturation for intact PTH is never reached neither in health nor in disease.

The situation is remarkably different for C-regional PTH. 'Distribution', generation and metabolism of C-regional PTH peptides are slowed down. From studies of *Silverman and Yalow* (21) we know that the secretory rate of C-regional peptides is small compared to intact PTH under conditions of uremia, indicating that peripheral metabolic rate is in fact altered. Whereas intact PTH is nearly adequately inactivated by enzymes at the target organs, C-regional peptides are not readily cleaved.

Summary

Distribution and metabolic degradation of bPTH was studied in man after infusion of 400 U bPTH and blood sampling up to 2 h. Disappearance rates of intact 1-84 bPTH, carboxyl- and amino-regional peptides were calculated for healthy subjects (n = 12); patients suffering from moderate (GFR 15–30 ml/min, n = 4) and advanced (GFR $\leqslant$ 10 ml/min, n = 36) chronic renal failure; and bilateral nephrectomized patients (n = 3). Two components with a rapid and a slow disappearance rate can be separated ('distribution', 'metabolism'). Half-lives are found to be in the range known from animal experiments with marked differences between intact 1-84 PTH and peptide fragments. Influence of impaired renal function on metabolic turnover rates of PTH is given.

References

1 *Arnaud, C.D.; Sizemore, G.W.; Oldham, S.B.; Fischer, J.A.; Tsao, H.S., and Littledike, E.T.:* Human parathyroid hormone. Glandular and secreted species. Am. J. Med. *50:* 630–638 (1971).

2 *Babb, A.L.; Farrell, P.C.; Krelli, D.A., and Scribner, B.H.:* Hemodialyzer evaluation by examination of solute molecular spectra. Trans. Am. Soc. artif. internal Organs *18:* 98–105 (1972).

3 *Canterbury, J.M.; Bricker, L.A.; Levey, G.S.; Kozlovskis, P.L.; Ruiz, E.; Zull, J.E., and Reiss, E.:* Metabolism of bovine parathyroid hormone. Immunological and biological characteristics of fragments generated by liver perfusion. J. clin. Invest. *55:* 1245–1253 (1975).

4 *Catherwood, B. and Singer, F.R.:* Generation of a carboxyl-terminal fragment of bovine parathyroid hormone by canine renal plasma membranes. Biochem. biophys. Res. Commun. *57:* 469–475 (1974).

5 *Goltzman, D.; Peytremann, A.; Callahan, E.N.; Segre, G.V., and Potts, J.T., jr.:* Metabolism and biological activity of parathyroid hormone in renal cortical membranes. J. clin. Invest. *57:* 8–19 (1976).

6 *Hehrmann, R.; Wilke, R.; Nordmeyer, J.P. und Hesch, R.D.:* Hochsensitiver, C-terminal-spezifischer Radioimmunoassay für menschliches Parathormon als Routinemethode. Dt. med. Wschr. *101:* 1726–1729 (1976).

7 *Hesch, R.D.; McIntosh, C.H.S., and Woodhead, J.S.:* New aspects of radioimmunochemical measurement of human parathyroid hormone using the labelled antibody technique. Horm. Metab. Res. *7:* 347–352 (1975).

8 *Hesch, R.D.; Ebel, H.; Hehrmann, R., and Jüppner, H.:* Endocrinological aspects of PTH metabolism in the kidney. Contr. Nephrol., vol. 13, pp. 104–114 (Karger, Basel 1978).

9 *Hruska, K.A.; Kopelman, R.; Rutherford, W.E.; Klahr, S., and Slatopolsky, E.:* Metabolism of immunoreactive parathyroid hormone in the dog. The role of the kidney and the effects of chronic renal disease. J. clin. Invest. *56:* 39–48 (1975).

10 *Jüppner, H.; Ebel, H.; Hehrmann, R., and Hesch, R.D.:* Characterization of the PTH inactivating enzyme (II). 6th Parathyroid Conf., poster III-04, Vancouver 1977.

11 *Kleeman, C.R.; Better, O.; Massry, S.G., and Maxwell, M.H.:* Divalent ion metabolism and osteodystrophy in chronic renal failure. Yale J. Biol. Med. *40:* 1–45 (1967).

12 *Martin, T.J.; Melick, R.A., and De Luise, M.:* Metabolism of parathyroid hormone. Degradation of radioiodinated hormone by a kidney enzyme. Biochem. J. *111:* 509–514 (1969).

13 *Martin, K.; Hruska, K.; Greenwalt, A.; Klahr, S., and Slatopolsky, E.:* Selective uptake of intact parathyroid hormone by the liver. Differences between hepatic and renal uptake. J. clin. Invest. *58:* 781–788 (1976).

14 *Massry, S.G.; Coburn, J.W.; Peacock, M., and Kleeman, C.R.:* Turnover of endogenous parathyroid hormone in uremic patients and those undergoing hemodialysis. Trans. Am. Soc. artif. internal Organs *18:* 416–421 (1972).

15 *Melick, R.A. and Martin, T.J.:* Parathyroid hormone metabolism in man. Effect of nephrectomy. Clin. Sci. *37:* 667–674 (1969).

16 *Nordquist, R.E. and Palmieri, G.M.A.:* Intracellular localization of parathyroid hormone in the kidney. Endocrinology *95:* 229–237 (1974).

17 *Okano, K.; Fujita, T.; Orimo, H.; Ohata, M., and Yoshikawa, M.:* Effect of renal injury on the activity of enzymes preferentially hydrolyzing parathyroid hormone. Endocr. jap. *18:* 327–333 (1971).

18 *Orimo, H.; Fujita, T.; Morii, H., and Nakao, K.:* Inactivation *in vitro* of parathyroid hormone activity by kidney slices. Endocrinology *76:* 255–258 (1965).

19 *Reiss, E. and Canterbury, J.M.:* Genesis of hyperparathyroidism. Am. J. Med. *50:* 679–685 (1971).

20 *Segre, G.V.; Niall, H.D.; Habener, J.F., and Potts, J.T., jr.:* Metabolism of parathyroid hormone. Physiologic and clinical significance. Am. J. Med. *56:* 774–784 (1974).

21 *Silverman, R. and Yalow, R.S.:* Heterogeneity of parathyroid hormone. Clinical and physiologic implications. J. clin. Invest. *52:* 1958–1971 (1973).

22 *Slatopolsky, E.; Hruska, K., and Rutherford, W.E.:* Current concepts of parathyroid hormone and vitamin D metabolism. Perturbations in chronic renal disease. Kidney int., Suppl., pp. 90–96 (1975).

23 *Strober, W. and Waldmann, T.A.:* The role of the kidney in the metabolism of plasma proteins. Proc. 6th int. Congr. Nephrol., Florence 1975, pp. 392–405 (Karger, Basel 1976).

24 *Zull, J.E.; Shriver, J., and Chuang, J.:* in *Norman, Schaefer, Grigoleit, Herrath and Ritz* Vitamin D and problems related to uremic bone disease, pp. 431–438 (de Gruyter, Berlin 1975).

Dr. *N. Lustenberger,* Departement für Innere Medizin, Medizinische Hochschule Hannover, Karl-Wiechert-Allee 9, *D–3000 Hannover 61* (FRG)

Contr. Nephrol., vol. 13, pp. 125–131 (Karger, Basel 1978)

Skeletal Resistance to the Calcemic Action of Parathyroid Hormone: Role of the Kidney and Vitamin D

Shaul G. Massry

Division of Nephrology and Department of Medicine, The University of Southern California School of Medicine, Los Angeles, Calif.

A series of studies from our laboratory have shown that patients with renal failure display impaired calcemic response to both exogenous and endogenous parathyroid hormone (PTH) (1–5). This skeletal resistance to the calcemic action of PTH has been demonstrated in patients (2) and animals (4, 5) with acute renal failure, patients with early (1, 3) and advanced renal insufficiency (1), dialysis patients (1) and renal transplant recipients with impaired renal function (1).

The mechanism(s) underlying this phenomenon is complicated and may be multifactorial. A rise in the level of serum calcium after the administration of PTH is primarily caused by the mobilization of bone calcium into the extracellular fluid. Several factors may modify this effect of PTH. These may include an increase in the endogenous levels of PTH, hyperphosphatemia, hypocalcemia, presence of osteomalacia, a deficiency of 1,25-dihydroxy-vitamin D ($1,25[OH]_2D_3$) and uremia, *per se.*

The infusion of PTH to uremic patients who already had elevated blood levels of PTH (6–8) may produce only a slight increment in the levels of the circulating hormone and, hence, cause a small calcemic response. Several lines of evidence vitiate this contention. First, the infusion of PTH to uremic patients after subtotal parathyroidectomy, when the circulating level of PTH was reduced toward normal, did not elicit a better calcemic response (1). Second, a 15- to 40-fold increment in the circulating levels of immunoreactive PTH, produced by PTH infusion, did not cause a normal calcemic response in patients with acute renal failure (2) (fig. 1). Third, the impaired calcemic response to PTH was demonstrated in parathyroidectomized dogs with acute renal failure produced by bilateral nephrectomy (4) (fig. 2).

Raisz (9) found that a low concentration of calcium ion in the media bathing culture of fetal rat bone can block the calcium mobilizing effect of PTH.

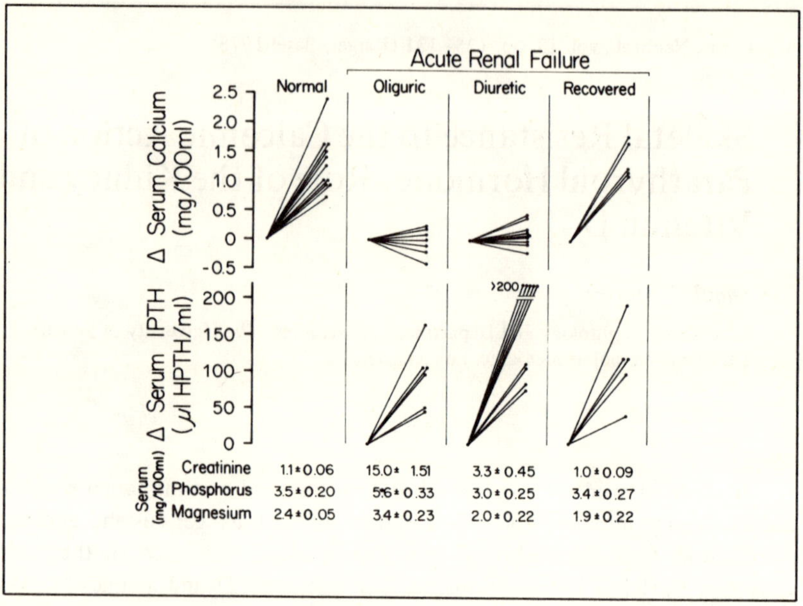

Fig. 1. The changes in the concentrations of serum calcium and in the blood levels of immunoreactive PTH during infusion of parathyroid extract in patients with acute renal failure. Based on data from *Massry et al.* (2).

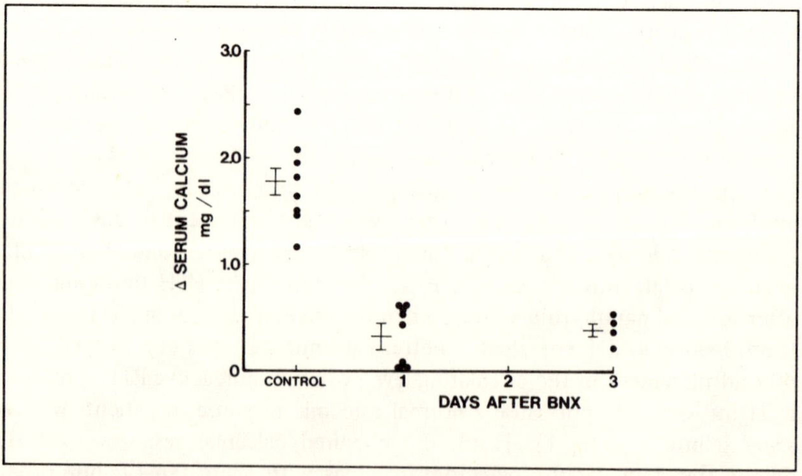

Fig. 2. The changes in the concentration of serum calcium produced by parathyroid extract infusion in thyroparathyroidectomized dogs before and after 1 and 3 days of bilateral nephrectomy (BNX). Each point represents one study. The brackets denote one standard error. From *Massry et al.* (4).

Evanson (10) reported that the calcemic response to PTH in uremic patients with serum calcium concentrations below 7.0 mg/100 ml was less than that observed in patients without hypocalcemia. In contrast, *Massry et al.* (1) have demonstrated that the changes in serum calcium during the infusion of PTH into patients with uremia were not related to the initial levels of serum calcium. The reason for the differences between the data of *Massry et al.* (1) and that of *Evanson* (10) might have been due to the presence of osteomalacia in *Evanson*'s patients. Indeed, 2 of the hypocalcemic patients studied by *Evanson* had pseudofractures and elevated plasma levels of alkaline phosphatase.

Hyperphosphatemia, which is usually present in patients with chronic renal failure, may interfere with the calcemic response to PTH. However, two observations from our laboratory militate against a major role for hyperphosphatemia in the genesis of this abnormality. First, there was no significant correlation between the blood levels of serum phosphorus and the changes in the concentrations of serum calcium produced by PTH infusion (1). Second, studies in patients with acute renal failure have demonstrated the presence of skeletal resistance to the calcemic action of PTH whether the blood levels of serum phosphorus were normal or high (2).

Osteomalacia may render the skeleton unresponsive to the calcemic action of PTH, presumably because of accumulation of osteoid over bone surfaces (11). Although osteomalacia may be present in patients with chronic renal failure (12), it is unlikely that excess osteoid accumulates in adequate amounts to account for the impaired calcemic response to PTH in patients or animals with acute renal failure. In one study of bone biopsy specimens from patients with acute renal failure, there was evidence for a mild increase in bone resorption but without increased osteoid (13).

It is evident that elevation in endogenous levels of PTH, hyperphosphatemia, hypocalcemia or accumulation of osteoid on bone surfaces could not provide adequate explanation for the impaired calcemic response to PTH in uremia. Theoretically, at least two additional possibilities should be considered.

First, it has been shown that the active metabolite $1{,}25(OH)_2D_3$ is produced by the kidney from the parent vitamin (14–16). A deficiency of $1{,}25(OH)_2D_3$ could, therefore, occur in patients with renal failure. Indeed, *Brumbaugh et al.* (17) reported very low concentrations of $1{,}25(OH)_2D_3$ in such patients. Since vitamin D may play a permissive role in the action of PTH on bone (18), it is possible that a deficiency in $1{,}25(OH)_2D_3$ is, partially or totally, responsible for the skeletal resistance to the calcemic action of PTH in uremia.

Second, uremia, *per se,* may interfere either with the production of $1{,}25(OH)_2D_3$ or with its action on target organ and, as such, contributes to the impaired calcemic response to PTH. Also, retention of some as yet unidentified toxic material during uremia may alter bone metabolism and underlie the defect in the response to PTH.

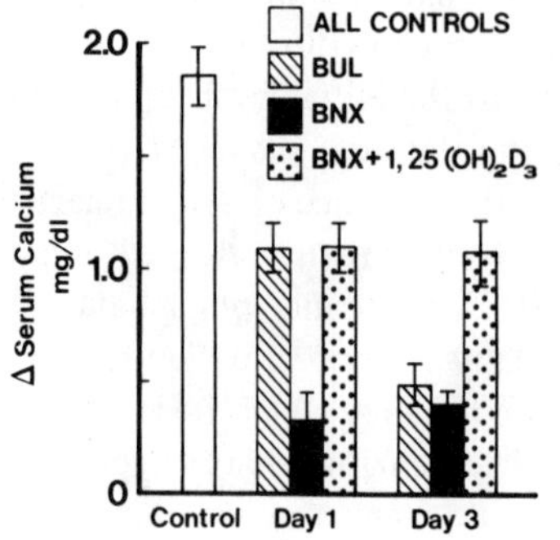

Fig. 3. The changes in the concentration of serum calcium produced by parathyroid extract infusion in thyroparathyroidectomized dogs before and after the induction of uremia. Uremia was produced either by bilateral ureteral ligation (BUL) and bilateral nephrectomy (BNX). Also the effects of $1,25(OH)_2D_3$ administration to nephrectomized dogs are presented.

Studies were carried out in our laboratory to investigate the role of $1,25(OH)_2D_3$ in the skeletal resistance to the calcemic action of PTH (19). The change in serum calcium after the intravenous infusion of 2 U parathyroid extract (PTE)/kg body weight/h for 8 h was evaluated in thyroparathyroidectomized dogs before and 1, 2 and 3 days after induction of uremia by bilateral ureteral ligation (11 dogs) or by bilateral nephrectomy (8 dogs). In another 6 nephrectomized and T-PTX dogs 0.68 μg of $1,25(OH)_2D_3$/day was given on day of nephrectomy and for 2 days thereafter. The results are shown in figure 3. Serum creatinine in each day of the study was not different among the three groups. The study also included the evaluation of the effect of sham operation (5 dogs) and the administration of $1,25(OH)_2D_3$ to dogs with normal renal function (4 dogs) on the calcemic response to PTE, as well as the reproducibility of such a response on the same animal. The results showed that (1) the calcemic response to PTE is markedly impaired after 1 day of bilateral ureteral ligation or nephrectomy, but the impairment is more severe after nephrectomy, (2) the calcemic response to PTE after 2 or 3 days of bilateral ureteral ligation was similar to that seen at 1 day after nephrectomy, (3) $1,25(OH)_2D_3$ partially restored the calcemic response to PTE in the nephrectomized animals to levels similar to those seen after 1 day of bilateral ureteral ligation, (4) sham operation did not affect the response to PTE and repeated infusion of PTE produced similar changes in the levels of serum calcium. The data are consistent with (a) a deficiency of $1,25(OH)_2D_3$ is, at least, partly responsible for the skeletal resistance to the calcemic action of PTH in uremia, (b) uremia, *per se,* may also contribute to this phenomenon and, (c) the kidney after 1 day of complete bilateral ureteral ligation may still produce $1,25(OH)_2D_3$, but this ability is compromised after 2 days of ureteral obstruction.

Additional studies from our laboratories evaluated the role of uremia, *per se,* in the genesis of the skeletal resistance to the calcemic action of PTH (20). The changes in serum calcium after infusion of 2 U PTE/kg body weight/h for 8 h were studied in thyroparathyroidectomized dogs before and after 1 and 3 days of acute uremia produced by diverting the ureters into a vein. These animals are uremic but have functioning kidneys. In another group of such uremic animals 0.68 μg of $1.25(OH)_2D_3$/day was given on day of surgery and for 2 days thereafter. The effect of reversal of uremia by disconnecting ureters from vein was also studied. Glomerular filtration rate during the uremic period remained similar to control. Results show that (1) calcemic response to PTE is markedly impaired after 1 day of uremia despite the presence of normally functioning kidney, (2) $1,25(OH)_2D_3$ completely restored the defect. This is in contrast to previous observations demonstrating partial correction of this abnormality in dogs with bilateral nephrectomy and similar uremia, (3) the response to PTE was normal after reversal of uremia. Data indicate that (1) uremia *per se,* contributes to the resistance to the calcemic action of PTE either by interfering with production of $1,25(OH)_2D_3$ or with its action on target organs, (2) this defect is reversible, and (3) $1,25(OH)_2D_3$ in uremic dogs with functioning kidneys is more effective than in nephrectomized dogs, suggesting the kidney may either produce a more polar substance of $1,25(OH)_2D_3$ and another vitamin D metabolite $24,25(OH)_2D_3$ act synergistically and, thus, are more effective. Preliminary studies from our laboratory indeed indicate that the administration of both $1,25(OH)_2D_3$ and $24,25(OH)_2D_3$ to bilaterally nephrectomized dogs restore the calcemic response to PTH to normal.

To further evaluate the mechanisms of the defective calcemic response to PTH, *Kurokawa et al.* (5) studied the effect of PTH on cyclic AMP of bone from uremic rats. These uremic animals displayed the defective calcemic response to PTH but the rise in bone cyclic AMP was similar to that observed in bones from nonuremic rats. These observations indicate that the defect in the calcemic response is not at the level of adenyl cyclase-cyclic AMP system of the bone, but at a subsequent step in the chain of cellular reactions induced by PTH. Indeed, recent studies from our laboratory demonstrated that the activation of cyclic AMP dependent protein kinase by PTH is defective in bones from uremic rats (21).

References

1 *Massry, S.G.; Coburn, J.W.; Lee, D.B.N.; Jowsey, J., and Kleeman, C.R.:* Skeletal resistance to parathyroid hormone in renal failure. Studies in 105 human subjects. Ann. intern. Med. *78:* 357–364 (1973).

2 *Massry, S.G.; Arieff, A.I.; Coburn, J.W.; Palmieri, G., and Kleeman, C.R.:* Divalent ion metabolism in patients with acute renal failure. Studies on mechanism of hypocalcemia. Kidney int. *5:* 437–445 (1974).

3 *Llach, F.; Massry, S.G.; Singer, F.R.; Kurokawa, K.; Kaye, J.H., and Coburn, J.W.:* Skeletal resistance to endogenous parathyroid hormone in patients with early renal failure. A possible cause for secondary hyperparathyroidism. J. clin. Endocr. Metab. *41:* 339–345 (1975).

4 *Massry, S.G.; Stein, R.; Garty, J.; Arieff, A.I.; Coburn, J.W.; Norman, A.W., and Friedler, R.M.:* Skeletal resistance to the calcemic action of parathyroid hormone in uremia. Role of $1,25(OH)_2D_3$. Kidney int. *9:* 467–474 (1976).

5 *Kurokawa, K.; Oda, Y.; Aznar, E., and Massry, S.G.:* Effects of parathyroid hormone (PTH) on bone cyclic AMP (cAMP) in acute renal failure in rats. Clin. Res. *24:* 363A (1976).

6 *Berson, S.A. and Yallow, R.S.:* Parathyroid hormone in the plasma in adenomatous hyperparathyroidism, uremia, and bronchogenic carcinoma. Science *154:* 907–909 (1966).

7 *Reiss, E. and Canterbury, J.M.:* Genesis of hyperparathyroidism. Am. J. Med. *50:* 679–685 (1971).

8 *Massry, S.G.; Coburn, J.W.; Peacock, M., and Kleeman, C.R.:* Turnover of endogenous parathyroid hormone in uremic patients and those undergoing hemodialysis. Trans. Am. Soc. artif. internal Organs *18:* 416–422 (1972).

9 *Raisz, L.G.:* Physiologic and pharmacologic regulation of bone resorption. New Engl. J. Med. *282:* 909–916 (1970).

10 *Evanson, J.M.:* The effect of the infusion of parathyroid extract in hypocalcemic states. Clin. Sci. *31:* 63–75 (1966).

11 *Jowsey, J.:* Calcium release from the skeleton of rachitic puppies. J. clin. Invest. *51:* 9–15 (1972).

12 *Massry, S.G. and Coburn, J.W.:* Divalent ion metabolism and renal osteodystrophy; in *Massry and Sellers* Clinical aspects of uremia and dialysis, pp. 304–387 (Thomas, Springfield 1976).

13 *Zech, P.; Meunier, P.; Beruard, M.; Edouard, C.; Pozet, N., and Traeger, J.:* Quantitative bone histology in acute renal failure. Eur. Dial. Transplant. Ass. *10:* 316–323 (1973).

14 *Fraser, D.R. and Kodicek, E.:* Unique biosynthesis by kidney of a biologically active vitamin D metabolite. Nature, Lond. *228:* 764–766 (1970).

15 *Gray, R.; Boyle, I., and DeLuca, H.F.:* Vitamin D metabolism. The role of the kidney tissue. Science *172:* 1232–1234 (1971).

16 *Norman, A.W.; Midgett, R.J.; Myrtle, J.F., and Nowicki, H.G.:* Studies on calciferol metabolism. I. Production of vitamin D metabolite 4B from 25-OH-cholecalciferol by kidney homogenates. Biochem. biophys. Res. Commun. *42:* 1082–1087 (1971).

17 *Brumbaugh, P.F.; Haussler, D.H.; Bressler, R., and Haussler, M.R.:* Radioreceptor assay for 1α,25-dihydroxyvitamin D_3. Science *183:* 1089–1091 (1974).

18 *Arnaud, C.; Rasmussen, H., and Anast, C.:* Further studies on the interrelationship between parathyroid hormone and vitamin D. J. clin. Invest. *45:* 1955–1964 (1966).

19 *Massry, S.G.; Stein, R.; Garty, J.; Arieff, A.I.; Coburn, J.W.; Norman, A.W., and*

Friedler, R.M.: Skeletal resistance to the calcemic action of parathyroid hormone in uremia. Role of 1,25$(OH)_2D_3$. Kidney int. *9:* 467–474 (1976).
20 *Massry, S.G.; Dua, S.; Garty, J., and Friedler, R.:* Role of uremia in the skeletal resistance to the calcemic action of parathyroid hormone (PTH). Proc. Am. Soc. Nephrol. *9:* 4 (1976).
21 *Oda, Y.; Kurokawa, K.; Aznar, E.; Kreusser, W.J., and Massry, S.G.:* Effects of parathyroid hormone (PTH) on bone cyclic (cAMP) and cAMP-dependent protein kinase activity (cAMP-PK) in acute renal failure in rats. Clin. Res. *25* (in press, 1978).

Shaul G. Massry, MD, Chief, Division of Nephrology, USC School of Medicine, 2025 Zonal Avenue, *Los Angeles, CA 90033* (USA)

Contr. Nephrol., vol. 13, pp. 132–146 (Karger, Basel 1978)

Vitamin D and the Syndromes of Azotaemic Osteodystrophy[1]

S. W. Stanbury

Department of Medicine, The Royal Infirmary, Manchester

In my first publication on the subject 20 years ago, I expressed the conviction that the development of azotaemic osteodystrophy was related in some way to lack of the biological effects of vitamin D; 'the rickets or osteomalacia of chronic uraemia results from an acquired insensitivity to the action of vitamin D' (44). This view was based primarily on the finding in many patients with chronic uraemia of defective mineralisation in the growth cartilage and the bone – changes that were indistinguishable from those produced by simple vitamin D deficiency (2). It was supported by preliminary studies which demonstrated impaired intestinal absorption of calcium in these patients and a curative response to pharmacological doses of vitamin D. None of this was actually new. In 1943, *Liu and Chu* (27) had studied mineral metabolism and the effect of vitamin D in similar patients and suggested that some factor in the uraemic state interfered with the action of vitamin D. Even 20 years before this, shrewd clinicans had carefully distinguished renal rickets from nutritional rickets and shown that the renal osteodystrophy could not be healed by cod liver oil (4, 40). These paediatricians in the 1920s were still seeing nutritional rickets in their everyday practise, an experience that qualified them particularly well to separate this disease from renal rickets. This is to be remembered in assessing recent claims that osteomalacia in renal failure is actually due to vitamin D deficiency (15, 17). In my opinion, these recent proposals are unacceptable (49).

Shortly after the discovery (21) of the role of the kidney in producing the hormonal metabolite of vitamin D, 1,25-dihydroxycholecalciferol, we were able to demonstrate that patients with chronic renal failure produced essentially no 1,25-dihydroxycholecalciferol from an injected pulse dose of radioactive cholecalciferol (37), and it has subsequently been proved by direct measurements of the sterol in the plasma that there is an absolute deficiency of this hormone in

[1] The investigations reported were supported by a grant from the British Medical Research Council.

renal failure (19, 22). In the light of these findings, it might be thought that little more can be said about the role of vitamin D in renal bone disease. But, reviewing one's cumulative experience of patients with bone disease complicating conservatively treated renal failure (45, 47, 52), it appears that problems delineated as long ago as 1962 (50) still remain unresolved. Many of these problems involve the role of vitamin D and include: (1) the mechanisms of action of large doses of the vitamin; (2) the stage of renal disease at which so-called 'vitamin D resistance' develops; (3) the question of a direct action of vitamin D on the mineralisation of bone; (4) the factors responsible for development of hypocalcaemia and hypercalcaemia in renal failure; (5) the role of vitamin D in the secondary hyperparathyroidism of renal failure. In an attempt to understand the role of vitamin D, independent of the many other factors operating in the patient with renal impairment, we have more recently studied about 150 adults and adolescents with simple vitamin D deficiency and normal renal function. Among these patients we have encountered most of the biochemical and osteological phenomena seen in renal osteodystrophy. This experience of the osteodystrophy of vitamin D deficiency provides a model that can be applied cautiously to the problems of renal osteodystrophy (53).

Vitamin D Resistance in Renal Failure; the Role of 25-Hydroxycalciferol in Mediating the Beneficial Therapeutic Effect of Pharmacological Doses of Calciferol

In investigating the phenomenon of 'resistance to vitamin D' in renal failure, we used the intestinal net absorption of calcium as an index of vitamin D action (50). Firstly, because calcium absorption could be measured fairly easily and, secondly, because an effect on intestinal calcium transport is universally acknowledged as a primary action of the vitamin. In a group of azotaemic patients, receiving a calcium intake of between 230 and 1,200 mg/day, it was established that the faecal calcium did not differ significantly from the dietary calcium (46). The intestinal net absorption of calcium was thus close to zero and, since the urinary output of calcium was low, the external calcium balance was close to equilibrium or negative in minor degree (46, 47). It was shown that this intestinal malabsorption of calcium could invariably be overcome by treatment with pharmacological (milligram) doses of ergocalciferol or dihydrotachysterol, and this was always followed by clinical healing of accompanying rickets or osteomalacia (50, 52). In the individual patient a daily dose of 2.5 mg of calciferol might prove ineffective, whereas 7.5 mg/day both promoted calcium absorption and effected clinical cure (50). Consequently a detailed investigation of this 'apparent vitamin D resistance' was made in 17 azotaemic patients. Progressively increasing doses of ergocalciferol were administered orally and the effects of this treatment observed on the external mineral balance and the total vitamin D activity in plasma – as measured by the biologically assayed anti-ricketic activity in plasma (28). It was found that promotion of calcium absorption in patients with chronic renal failure appeared to require a greater intake of the vitamin and higher levels of 'vitamin D activity' in the plasma than were required by healthy individuals (28).

Unfortunately, when these extremely tedious investigations were done it was not possible to measure the serum concentration of 25-hydroxycalciferol. If it were assumed that all the biologically assayed activity in plasma were due to this metabolite, the observations made (28) might imply that concentrations of 25-hydroxycalciferol as high as 100–250 μg/l could fail to increase calcium absorption. In fact, this assumption was not justified as we were later able to demonstrate that during treatment with milligram doses of calciferol a significant proportion of the total 'vitamin D activity' in plasma is attributable to unmetabolised vitamin (38). The range of concentration of 25-hydroxycalciferol in plasma that is required to promote calcium absorption in the uraemic or anephric patient is unknown; nor is it known if this quantity varies from patient to patient or at different stages of renal failure. Data recently published by *Eastwood et al.* (17) provide an initial estimate of this requirement. In 2 patients with azotaemic osteomalacia receiving 20 μg of 25-hydroxycholecalciferol *intravenously* per day, no effect on calcium absorption or on the mineralisation front in bone was observed with plasma 25-hydroxycalciferol levels of 37 and 40 μg/1; both indices of response were observed in 3 other patients (receiving up to 40 μg/day of the sterol) with the plasma-25-hydroxycalciferol between 63 and 80 μg/l. The authors interpreted their results to imply that the osteomalacia in these patients – and in azotaemic patients generally – was due to vitamin D deficiency. Unfortunately, the amount of administered 25-hydroxycholecalciferol required to correct simple vitamin D deficiency in man is unknown, but we have seen evidence of a curative response when the serum 25-hydroxycalciferol has increased from 2 to no more than 5 μg/l. Elsewhere, we have estimated that perhaps as little as 0.2–0.5 μg/day of intravenously administered 25-hydroxycholecalciferol might suffice to correct simple vitamin D deficiency in man (49). Direct comparison of patients with renal failure and those with simple vitamin D deficiency is not feasible, since the latter have a normal capacity to produce the requisite picomolar amounts of the hormonal effector, 1,25-dihydroxycholecalciferol. Nonetheless, the doses of 25-hydroxycholecalciferol given by *Eastwood et al.* (17) may be 100 times the normal requirement. It is of interest that the binding protein extractable from the intestinal mucosal nuclei of the vitamin D deficient chick has a binding affinity for 25-hydroxycholecalciferol about one hundredth of that for 1,25-dihydroxycholecalciferol (*C.M. Taylor*, to be published).

All the evidence indicates that the intestinal malabsorption of calcium in patients with renal failure is due to a deficiency of 1,25-dihydroxycholecalciferol (37). This is supported by animal experiments that demonstrated absence of this sterol in the intestinal mucosa of anephric rats (25a) and a quantitative reduction of its mucosal concentration in rats submitted to ureteric ligation (25a) or to 5/6 nephrectomy (24). In all these experiments, the capacity of the intestinal mucosa to transport calcium *in vitro* correlated with its content of 1,25-dihydroxycholecalciferol. At the same time, both in uraemic man and uraemic animals, the hepatic capacity to 25-hydroxylate vitamin D remained normal; despite continued claims to the contrary there is no evidence for an abnormality of the hepatic phase of vitamin D metabolism in uraemic man (36–38). During treatment with pharmacological doses of vitamin D, the serum 25-hydroxycalciferol may reach very high levels and it must be assumed that this excess of the 25-hydroxylated derivative can substitute at receptor sites for the deficient 1,25-dihydroxy vitamin D (48).

Plasma [Ca] × [P] Product and Defective Mineralisation in Renal Failure; the Problem of a Direct Effect of Vitamin D on Mineralisation of Osteoid

Our earliest studies suggested that defective skeletal mineralisation in renal failure could not be related causally to the plasma [Ca] × [P] product (50). In patients with untreated azotaemic rickets or osteomalacia this product could vary from 18 to 67. Treatment with doses of vitamin D, which were effective in producing intestinal absorption and bodily retention of calcium and phosphorus, always caused some increase in the serum calcium or serum P_i with an increase in the [Ca] × [P] product; but these changes were small and no particular value of the product was associated with a curative response (50). Subsequently it was possible to demonstrate mineral retention during treatment with vitamin D without accompanying change, or even with a reduction, of the plasma [Ca] × [P] product (28, 54). It was evident that neither the *presence* of osteomalacia nor its putative *healing* during treatment with vitamin D was related predictably to the plasma [Ca] × [P] product, and it was concluded that vitamin D must facilitate the mineralisation of osteoid by a direct action on the bone (45).

This point could not be established with certainty, since the occurrence of bony mineralisation during treatment with vitamin D was inferred indirectly from the documentation of retention of calcium and phosphorus in a ratio (∿ 2:1 by weight) appropriate to deposition in bone. This inference was vindicated by the subsequent clinical and radiographic healing of the bone disease but relatively few observations were made on the effects of vitamin D treatment on the bone histology (45). *Bordier et al.* (6) showed that administration of vitamin D to patients with the osteomalacia of vitamin D deficiency increased the extent of the calcification front in bone. This effect could not be reproduced by induced elevation of the serum P_i in the absence of vitamin D treatment (10) and these collective observations were also interpreted to indicate a direct action of 'vitamin D' on bone. *Eastwood et al.* (16b) utilised measurements of the extent of the calcification front in bone to analyse the mechanism of action of vitamin D in azotaemic osteomalacia. They showed conclusively that elevation of the plasma [Ca] × [P] product, by simultaneous treatment with oral calcium carbonate and intravenous sodium phosphate, produced no increase in the calcification front although it caused some patchy mineralisation of osteoid. Conversely, treatment with vitamin D produced a highly significant extension of the calcification front, even when elevation of the plasma [Ca] ×[P] product was prevented or its level simultaneously reduced. These observations go as far as is possible in clinical studies towards 'proving' that vitamin D has a direct effect on osseous mineralisation in azotaemic osteomalacia. The insuperable difficulty in these studies, as in our own, is in excluding undetected post-prandial increases of serum calcium and serum P_i. Thus, during treatment with vitamin D in patients with primary hyperparathyroidism complicated by vitamin D deficiency, we find that there may be a difference of 1–1.5 mg/dl in the concentration of serum calcium between blood samples obtained in the fasting and non-fasting state (*Davies and Stanbury,* unpublished). Nonetheless, the only *reasonable* interpretation of these various observations is that 'vitamin D' does have a direct role in facilitating mineralisation and that lack of this effect contributes to the osteomalacia of renal failure.

The fact that administered 25-hydroxycholecalciferol (3), or 25-hydroxycalciferol produced metabolically from administered vitamin D, can heal azotaemic osteomalacia does not indicate that this sterol has a physiological direct role in promoting mineralisation of osteoid (7, 8) nor that a deficiency of 25-hydroxycalciferol is responsible for the osteomalacia of renal failure (15, 17); nor does the demonstration in bone cells of a cytosolic binding protein with a high affinity for 25-hydroxycholecalciferol (18). Reasons have been given elsewhere for rejecting the possibility that dihydroxylated derivatives other than 1,25-dihydroxycalciferol – that is either 24,25- or 25,26-dihydroxycalciferol – are responsible for the beneficial effects of calciferol in renal failure (49). It has been demonstrated that small (0.5–1.35 μg/day) intravenous doses of 1,25-dihydroxycholecalciferol will increase the front of calcification in azotaemic osteomalacia (16a) and many publications attest to the therapeutic effectiveness of continued treatment with dosage of that order (11–14, 23, 42). Although it is probable that the various therapeutic regimens utilising 1,25-dihydroxycholecalciferol, or its synthetic analogue 1α-hydroxycholecalciferol, have not yet reproduced physiological conditions (see below and fig. 1), the information available to date is compatible with a deficiency of 1,25-dihydroxycalciferol being responsible for the osteomalacia as well as the intestinal malabsorption of calcium in renal failure.

The alternative view that 25-hydroxycholecalciferol is responsible directly for mineralisation of osteoid is based principally on two observations. *Firstly,* that osteomalacia may be absent in some anephric patients (9), circumstances in which there can be no formation of 1,25-dihydroxycalciferol. But the obligatory dialytic therapy in such patients could enforce a positive mineral balance and healing of the osteomalacia, even if in an unphysiological manner. Healing of vitamin D deficiency osteomalacia has been produced by enforced retention of calcium and phosphorus without correcting the original vitamin deficiency (41). It is not known if *complete* healing produced in this manner would also restore a calcification front; nor is it known if vitamin D deficiency *without osteomalacia* is associated with reduction of the extent of the calcification front. *Secondly,* comparisons have been made of the effects of administered cholecalciferol, 25-hydroxycholecalciferol and 1,25-dihydroxycholecalciferol in vitamin D deficiency osteomalacia; and it is reported that 1,25-dihydroxycholecalciferol is least effective in increasing the extent of the calcification front (7, 8). Comparisons of this kind can be interpreted only if it is known precisely what the individual therapy is doing. Administration of vitamin D or 25-hydroxycholecalciferol will produce sustained elevation of the serum 25-hydroxycalciferol from which 1,25-dihydroxycalciferol will be produced in a smoothly regulated fashion. An oral dose of 1 μg of 1,25-dihydroxycholecalciferol produces only a transient elevation of its serum concentration, which barely reaches normal values (35) (fig. 1). That same dose may, however, expose the intestinal mucosa to a concentration of hormone up to two orders greater than physiological and will certainly activate intestinal transport of calcium; but the circulating concentration of hormone could be inadequate to saturate receptors in bone, parathyroid glands etc. Thus a valid comparison of the effects on the calcification front of different vitamin D sterols would necessitate the maintenance of physiological levels in the serum during the relevant period of observation; that is, the serum 1,25-dihydroxycholecalciferol should be maintained at concentrations of about 30–35 ng/l (fig. 1). A similar target would be the ideal for effective replacement therapy in renal failure and it is evident that contemporary therapeutic regimens fail to attain that ideal.

Pleomorphic Plasma Biochemistry in Azotaemic Osteodystrophy: the Problem of Hypocalcaemia and Hypercalcaemia; the Syndrome of Relative Hypoparathyroidism or Acquired 'Pseudohypoparathyroidism'

In early studies, one of the major problems in attempting to impose some sort of order on the chaos of azotaemic osteodystrophy was the extraordinarily wide variation in the concentrations of serum calcium and serum P_i in different affected patients. In a large collection of cases the serum calcium varied from 4.8

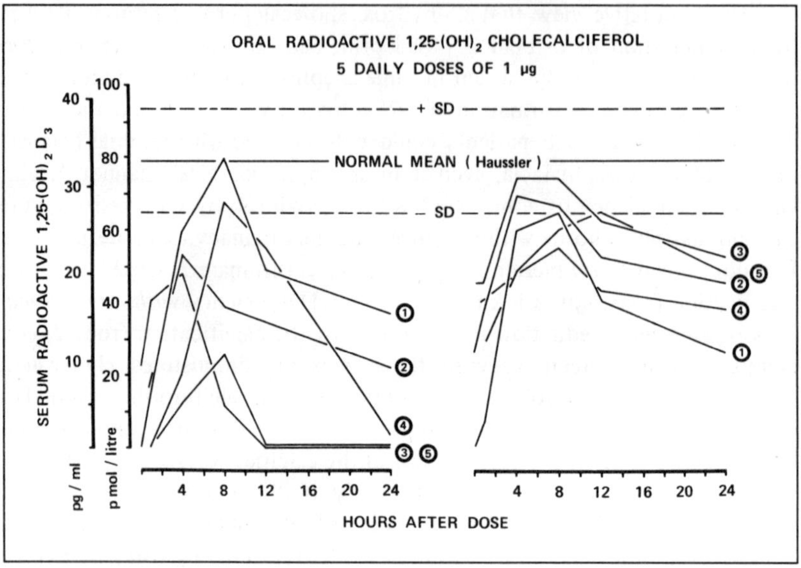

Fig. 1. The effects on the serum concentration of 1,25-dihydroxycholecalciferol of administering daily 1-µg doses of the radioactive hormone to 2 patients. From published (35) and unpublished data of Dr. *E.B. Mawer.* The mean and SD in normal subjects are taken from data of *Hughes et al.* (25b).

to 13.6 mg/dl, the serum P_i from 2.8 to 17.3 mg/dl and the [Ca] × [P] product from 20 to 200 $(mg/dl)^2$ (51). As a first approximation it was shown that azotaemic rickets and osteomalacia were statistically (but not individually) hypocalcaemic states and that renal generalised osteitis fibrosa was statistically associated with normocalcaemia. It was argued that this difference in serum calcium levels was a function of the effectiveness of the parathyroid glands in compensating for the effects of the 'apparent vitamin D deficiency' of renal failure (51). Although this analysis was unsupported by measurements of serum iPTH (immunoreactive parathyroid hormone) it accorded with available measurements of parathyroid mass. This interpretation, as it applies to conservatively managed renal failure, has generally survived the test of time and some confirmation has been provided by determinations of serum iPTH (1). Some physicians have also attempted to apply this analysis of serum biochemical findings to the osteodystrophy in dialysed patients; this is invalid, since the blood chemistry in such patients is as much determined by therapy as by endogenous homeostatic mechanisms.

Among the many problems remaining after this initial and relatively simple analysis, two were of especial interest. Firstly, the nature of the mechanism responsible for causing hypocalcaemia in these patients – particularly the occurrence of profound hypocalcaemia in patients without especially great elevation of the serum P_i. Secondly, the existence of a small but significant proportion of patients in whom the serum calcium was raised above normal, in spite of a greatly elevated serum P_i in many of them.

Clinical experience of a large number of patients indicated that stable hypocalcaemia could not be explained by elevation of the serum P_i, and any contribution from the impaired intestinal absorption of calcium appeared to be of relativel minor importance (45, 50). *Evanson* (20) demonstrated that the 'calcaemic' effect of infused parathyroid extract was less than normal in patients with renal failure. Thus, although most such patients showed evidence of parathyroid action on bone in the form of increased osteoclasis and osteitis fibrosa, the independent action of the hormone in raising the serum calcium was impaired. Using a similar technique, *Massry et al.* (32) studied the phenomenon systematically and showed that parathyroid extract produced a lesser ($\sim$ 36% of controls) elevation of the serum calcium in patients with advanced renal failure than in normal subjects. The same resistance to parathyroid extract was present in patients with moderate renal impairment (32) and also in acute renal failure (31). In a subsequent experimental exploration of the problem, it was shown that 1,25-dihydroxycholecalciferol would partially restore responsiveness to parathyroid extract in dogs rendered acutely uraemic by bilateral nephrectomy (34) and restore a completely normal response in acute uraemia produced by diverting the ureters into a vein (33). The full story is not yet elucidated and, as indicated previously (45), in chronic uraemia in man the presence of osteomalacia itself may contribute to the parathyroid unresponsiveness of bone. None the less, it is evident that lack of an action of 1,25-dihydroxycholecalciferol on bone contributes to the hypocalcaemia of renal failure. The demonstration of 'calcaemic unreponsiveness' in patients with moderate renal impairment (32) is of especial interest in view of the recent demonstration that the histological changes of osteomalacia may develop in the earliest stages of renal disease (5, 30). Taken together, these different observations suggest a need to seek for evidence of impaired synthesis of 1,25-dihydroxycholecalciferol in early renal disease.

The occurrence of hypercalcaemia in azotaemic patients with renal generalised osteitis fibrosa was originally attributed to a *qualitative* change in parathyroid function, it being assumed that prolonged stimulation of the glands had led to the development of autonomy of function (55). This was based partly on the nature of the accompanying bony changes and on the finding of apparently adenomatous transformation in surgically resected hyperplastic glands (55); it has also been shown subsequently that the serum iPTH in such patients is exceptionally high (1). With increasing experience it became apparent that the parathyroid glands in these patients usually showed only chief cell hyperplasia without adenomatous transformation and it was suggested alternatively that the apparent functional autonomy was due to the mass of hyperplastic tissue present (47). This hypothesis has been validated by clinical experiment (43) and it has even been possible to reverse the hypercalcaemic syndrome by therapeutic measures causing prolonged suppression of parathyroid function (57). Thus it is questionable whether so-called 'tertiary hyperparathyroidism' exists and, if it does, it must be extremely rare. The unresolved problem is to explain why parathyroid hyperplasia should proceed to a degree sufficient to cause hypercalcaemia.

We undertook studies in human vitamin D deficiency to seek analogies between that condition and the deficiency of 1,25-dihydroxycholecalciferol in renal failure; some of the analogies found were surprisingly close. Thus, the serum calcium in vitamin D deficiency varied from the grossly subnormal to actual hypercalcaemic levels; and, although the serum P_i was usually reduced, it could be increased above normal (53, 56). The full implications of this study cannot be discussed here and several aspects of the investigations have been published (29, 53, 56), but one group of observations might be mentioned as probably relevant to renal failure. In some adolescents with clinical nutritional rickets and profound hypocalcaemia ($\leqslant$7 mg/dl), the serum P_i was at the upper extent of normal or was actually increased (6–7 mg/dl), a biochemical syndrome not infrequent in azotaemic osteodystrophy (45). It was found that this relative hyperphosphataemia was resistant to the phosphaturic effect of infused parathyroid extract, and since the serum iPTH was increased in these patients, this syndrome could be classified as a form of acquired pseudohypoparathyroidism (56). Further analysis indicated that the severe hypocalcaemia in some of these patients was directly attributable to vitamin D deficiency; in others it was due to *correction* of the vitamin D deficiency and consequential deposition of calcium in the healing bone. Thus it appeared to be the severe hypocalcaemia itself that was responsible for the phosphaturic unresponsiveness to parathyroid extract, and the operation of a similar mechanism must be considered in chronically azotaemic patients with comparable degrees of hypocalcaemia.

Role of Vitamin D in Secondary Hyperparathyroidism of Renal Failure

It is, however, in relation to the secondary hyperparathyroidism of renal failure that analogies with simple vitamin D deficiency are of greatest interest. From studies of azotaemic osteodystrophy we have always felt that the effects of 'apparent vitamin D deficiency' – the intestinal malabsorption of calcium and especially the 'calcaemic unresponsiveness' of bone – was of greatest significance in causing secondary hyperparathyroidism (45, 50). The important contribution of 'phosphate retention' in influencing secondary hyperparathyroidism, as delineated by the classical studies of *Bricker* and *Slatopolsky*, are fully accepted (47,

48), but the personal feeling persists that this may be of secondary importance. It is difficult to explain the reversal of secondary hyperparathyroidism during treatment with vitamin D, despite persisting or increasing hyperphosphataemia, except in terms of the action of 'vitamin D' (45, 47, 50), and it has now been shown directly that treatment with 1,25-dihydroxycholecalciferol will reduce the serum iPTH in azotaemic osteodystrophy irrespective of the prevailing level of serum P_i (12, 14).

In simple vitamin D deficiency the static and kinetic responses of the parathyroid glands can be studied uncomplicated by the renal retention of hormonal fragments which bedevils studies in renal failure. Results of such studies can only be catalogued here. It is the rule to find an elevated serum iPTH in simple vitamin D deficiency and its concentration can be up to 25 times the average normal value (29, 53, 56). In patients with the highest levels of serum iPTH the bony changes – with osteitis fibrosa, marrow fibrosis and formation of woven bone – may be indistinguishable from severe azotaemic osteodystrophy. Thus chronic vitamin D deficiency in patients with normal renal function can produce as severe a secondary hyperparathyroidism as is encountered in renal failure. Parenthetically, in relation to the Bricker-Slatopolsky mechanism, ricketic patients with profound hypocalcaemia and hyperphosphataemia have lower levels of serum iPTH than do normocalcaemic patients (56). In vitamin D deficient patients with serum calcium between $\sim$ 8 and $\sim$ 11 mg/dl, a barely detectable induced rise of serum calcium causes an apparent arrest of parathyroid secretion; in patients with profound hypocalcaemia (serum calcium $\leqslant$7 mg/dl) parathyroid secretion appears to be indifferent to small changes of serum calcium (53, 56). Thus the relationship between serum calcium and serum iPTH appears not to be linear and proportionate but can be best described as a sigmoid curve (fig. 2), as already shown in the cow (39). All these patterns of parathyroid response in vitamin D deficiency appear to be determined by changes of calcium concentration in the plasma or in the parathyroid glands; 'vitamin D' (or one of its metabolites) appears not to be involved in these responses but this possibility cannot be certainly excluded. The exceptional vitamin D deficient patient with moderate hypercalcaemia responds to a minute further rise in serum calcium with an apparent arrest of parathyroid secretion (29) as in normocalcaemic patients (53). Thus, despite the occurrence of hypercalcaemia, the parathyroid glands in these patients were not functioning autonomously. But after correction of vitamin D deficiency in the same patients, there is a very slow ($t^1/_2$ $\sim$ 6 months) reduction of the static concentration of serum iPTH which is only adequately explained by structural involution of massively hyperplastic parathyroid glands (29). These and certain other observations have been taken to imply that 'vitamin D' may have a direct action on the parathyroid glands (29). At present this is little more than speculation but the recent demonstration in parathyroid tissue of a cytosolic binding protein with a high affinity for 1,25-dihydroxycholecalciferol (26)

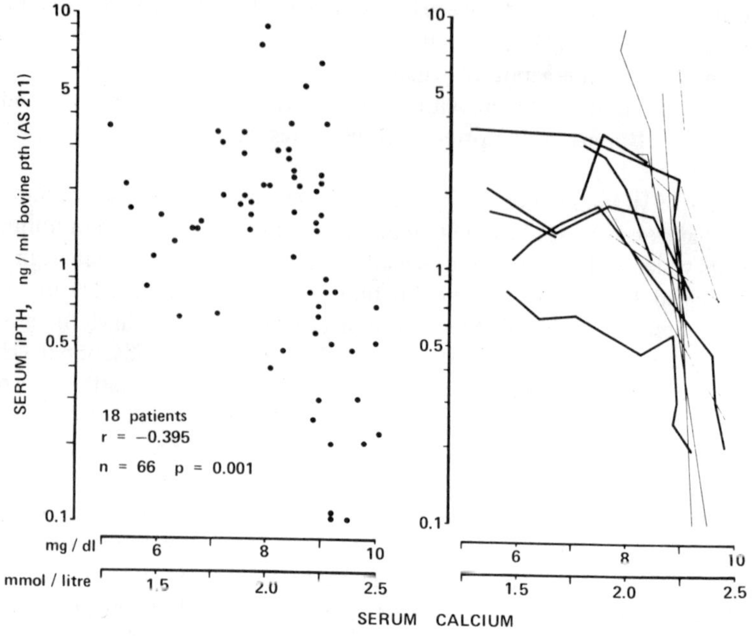

Fig. 2. The curvilinear relationship between serum calcium and serum iPTH in 18 patients before and after correction of simple vitamin D deficiency (56). Left: the scatter of 66 static measurements. Right: linear plots of sequential data from individual patients. Note: (1) in patients with initial serum calcium ⩾8 mg/dl (thin lines), a small increase of serum calcium during treatment causes an abrupt reduction of serum iPTH; (2) in profoundly hypocalcaemic patients (serum calcium 5–7 mg/dl) the serum iPTH is not significantly influenced by change of serum calcium until the latter reaches ∿ 8 mg/dl in the course of treatment. (Preliminary studies suggest a similar relationship in patients with renal failure.)

is compatible with the idea. If 1,25-dihydroxycholecalciferol did have a direct action on the parathyroid glands, its absence in vitamin D deficiency and in renal failure might account for several unexplained aspects of parathyroid function in these analogous conditions.

This has been a personal account, deliberately biassed to emphasise the role of vitamin D in azotaemic osteodystrophy. That there are many other factors contributing to the production of renal bone disease is beyond question but it still seems that 'lack of the biological effects of vitamin D' (50) is the most important single factor in its pathogenesis. This is intended to imply a quantitative deficiency of its hormonal metabolite, 1,25-dihydroxycholecalciferol; it has yet to be proved whether deficiency of 25-hydroxycholecalciferol or of 24,25-dihydroxycholecalciferol has any independent or collateral role.

Summary

Intestinal malabsorption of calcium and the development of osteomalacia in conservatively treated renal failure is explained by a quantitative deficiency of 1,25-dihydroxycholecalciferol, which also contributes to the development of hypocalcaemia. Excess of 25-hydroxycholecalciferol can substitute for this deficiency. The presence and healing of azotaemic osteomalacia is unrelated to the prevailing plasma [Ca] × [P] product. The data suggest that 'vitamin D' acts directly on bone mineralisation, but the claim that this apparent effect is normally due to 25-hydroxycholecalciferol is considered unproven. Most of the phenomena of azotaemic osteodystrophy are encountered in simple vitamin D deficiency; as in that condition, deficiency of 1,25-dihydroxycholecalciferol may be of primary singificance in causing secondary hyperparathyroidism in renal failure.

References

1 *Arnaud, C.D.:* Hyperparathyroidism in renal failure. Kidney int. *6:* 89–95 (1973).
2 *Ball, J.:* Diseases of bone; in *Harrison* Recent advances in pathology, pp. 293–338 (Churchill, London 1960).
3 *Balsan, S. and Witmer, G.:* The treatment of renal osteodystrophy with 25-hydroxycholecalciferol. 2nd Int. Symp. Pediat. Nephrology (Sandoz, Paris 1971).
4 *Barber, H.:* Renal dwarfism. A study of the course of the disease from seventeen cases. Guy's Hosp. Rep. *76:* 307 (1926).
5 *Bonucci, E.; Maschio, G.; D'Angelo, A.; Ossi, E.; Lupo, A., and Valvo, E.:* Morphological aspects of bone tissue in chronic renal disease. A histological and electron microscopic study; in *Norman, Schaefer, Grigoleit, von Herrath and Ritz* Vitamin D and problems related to uremic bone disease, pp. 523–530 (de Gruyter, Berlin 1975).
6 *Bordier, P.; Hioco, D.; Rouquier, M.; Hepner, G.W., and Thompson, G.R.:* Effects of intravenous vitamin D on bone and phosphate metabolism in osteomalacia. Calcif. Tiss. Res. *4:* 78–83 (1969).
7 *Bordier, P.; Marie, P.; Arnaud, C.D.; Gueris, J.; Ferriere, C., and Norman, A.W.:* Early effects of vitamin D and some analogues; $25OHD_3$; $1{,}25(OH)_2D_3$; 1α-$(OH)D_3$ – upon resorption formation and mineralisation of bone, in nutritional or malabsorption osteomalacia – with reference to serum iPTH, calcium, phosphate and alkaline phosphatase; in *Norman, Schaefer, Grigoleit, von Herrath and Ritz* Vitamin D and problems related to uremic bone disease, pp. 135–148 (de Gruyter, Berlin 1975).
8 *Bordier, P.; Marie, P.; Miravet, L.; Gueris, J.; Ferriere, C.; Ryckewaert, A., and Norman, A.W.:* Vitamin D deficiency-induced osteomalacia; contrasting effects of 25-$(OH)D_3$, $1{,}25(OH)_2D_3$ and $24{,}25(OH)_2D_3$ upon osteoid mineralisation; in Third Workshop in vitamin D (in press, 1977).
9 *Bordier, P.; Tun Chot, S.; Eastwood, J.B.; Fournier, A., and Wardener, H.E. de:* Lack of histological evidence of vitamin D abnormality in the bones of anephric patients. Clin. Sci. *44:* 33–41 (1973).

10 *Bordier, P.; Tun Chot, S.; Martin, J.; Queillé, M.L. et Hioco, D.:* Minéralisation du tissu ostéoïde chez l'ostéomalacique induite par surcharge phosphorée ou par vitamine D_3; dans *Hioco* Phosphate et métabolisme phosphocalcique, pp. 79–88 (Sandoz, Paris 1971).
11 *Brickman, A.S.; Coburn, J.W.; Massry, S.G., and Norman, A.W.:* 1,25-Dihydroxy-vitamin D in normal man and patients with renal failure. Ann. intern. Med. *80:* 161–168 (1974).
12 *Brickman, A.S.; Jowsey, J.; Sherrard, D.J.; Friedman, G.; Singer, F.R.; Baylink, D.J.; Maloney, N.; Massry, S.G.; Norman, A.W., and Coburn, J.W.:* Therapy with 1,25-dihydroxy-vitamin D in the management of renal osteodystrophy; in *Norman, Schaefer, Grigoleit, von Herrath and Ritz* Vitamin D and problems related to uremic bone disease, pp. 241–248 (de Gruyter, Berlin 1975).
13 *Brickman, A.S.; Coburn, J.W., and Norman, A.W.:* Action of 1,25-dihydroxycholecalciferol, a potent kidney-produced metabolite of vitamin D_3 in uremic man. New Engl. J. Med. *287:* 891–895 (1972).
14 *Coburn, J.W.; Brickman, A.S.; Sherrard, D.J.; Singer, F.R.; Baylink, D.J.; Massry, S.G., and Norman, A.W.:* Clinical efficacy of 1,25-$(OH)_2D_3$ in chronic renal failure; in Third Workshop on vitamin D (in press, 1977).
15 *Eastwood, J.B.; Harris, E.; Stamp, T.C.B., and Wardener, H.E. de:* Bone mineralisation and plasma 25-OH vitamin D in chronic renal failure. Lancet *ii:* 1209–1211 (1976).
16a *Eastwood, J.B.; Phillips, M.E.; Wardener, H.E. de; Bordier, P.J.; Marie, P.; Arnaud, C.D., and Norman, A.W.:* Biochemical and histological effects of 1,25-dihydroxycholecalciferol in the osteomalacia of chronic renal failure; in *Norman, Schaefer, Grigoleit, von Herrath and Ritz* Vitamin D and problems related to uremic bone disease, pp. 595–602 (de Gruyter, Berlin 1975).
16b *Eastwood, J.B.; Bordier, P.J.; Clarkson, E.M.; Tun Chot, S., and Wardener, H.E. de:* The contrasting effects on bone histology of vitamin D and of calcium carbonate in the osteomalacia of chronic renal failure. Clin. Sci. molec. Med. *47:* 23–42 (1974).
17 *Eastwood, J.B.; Stamp, T.C.P.; Wardener, H.E. de; Bordier, P., and Arnaud, C.D.:* The effect of 25-hydroxy vitamin D_3 in the osteomalacia of chronic renal failure. Clin. Sci. molec. Med. *52:* 499–508 (1977).
18 *Edelstein, S.:* Vitamin D-binding proteins. Biochem. Soc. spec. Publ. *3:* 43–54 (1974).
19 *Eisman, J.A.; Hamstra, A.J.; Kream, B.E., and De Luca, H.F.:* A sensitive, precise and convenient method for determination of 1,25-dihydroxy vitamin D in human plasma. Archs Biochem. Biophys. *176:* 235–243 (1976).
20 *Evanson, J.M.:* The response to the infusion of parathyroid extracts in hypocalcaemic states. Clin. Sci. *31:* 53–75 (1966).
21 *Fraser, D.R. and Kodicek, E.:* Unique biosynthesis by kidney of a biologically active vitamin D metabolite. Nature, Lond. *228:* 764–766 (1970).
22 *Haussler, M.R.; Baylink, D.J.; Hughes, M.R.; Brumbaugh, P.F.; Wergedal, J.E.; Shen, F.H.; Neilsen, R.L.; Counts, S.J.; Bursac, K.M., and McCain, T.A.:* The assay of 1α-25-dihydroxyvitamin D_3; physiologic and pathologic modulation of circulating hormone levels. Clin. Endocr. *5:* suppl., pp. 151–165 (1976).
23 *Henderson, R.G.; Russell, R.G.G.; Ledingham, J.G.G.; Smith, R.; Oliver, D.O.; Walton, R.J.; Small, D.G.; Preston, C.; Warner, G.T., and Norman, A.W.:* Effects of 1,25-dihydroxycholecalciferol on calcium absorption, muscle weakness and bone disease in chronic renal failure. Lancet *i:* 379–384 (1974).
24 *Hill, L.F. and Stanbury, S.W.:* Vitamin D and the kidney. Nephron *15:* 369–386 (1975).
25a *Hill, L.F.; Berg, C.J. van den, and Mawer, E.B.:* Vitamin D metabolism in experimental

uraemia: effects on intestinal transport of ^{45}Ca and on formation of 1,25-dihydroxy-cholecalciferol in the rat. Nature new Biol. *32:* 189–191 (1971).
25b *Hughes, M.R.; Baylink, D.J.; Jones, P.J., and Haussler, M.R.:* Radioligand receptor assay for 25-hydroxy-vitamin D_2/D_3 and 1α,25-dihydroxyvitamin D_2/D_3. Application to hypervitaminosis D. J. clin. Invest. *58:* 61–70 (1976).
26 *Hughes, M.R.; Zerwekh, J.E.; Brumbaugh, P.F., and Haussler, M.R.:* Mechanism of action of 1,25-dihydroxy vitamin D_3 at the intestine and parathyroid; in Third Workshop on vitamin D (in press, 1977).
27 *Liu, S.H. and Chu, H.I.:* Studies of calcium and phosphorus metabolism with special reference to pathogenesis and effect of dihydrotachysterol (DHT) and iron. Medicine, Baltimore *22:* 103–161 (1943).
28 *Lumb, G.A.; Mawer, E.B., and Stanbury, S.W.:* The apparent vitamin D resistance of chronic renal failure. A study of the physiology of vitamin D in man. Am. J. Med. *50:* 421–441 (1971).
29 *Lumb, G.A. and Stanbury, S.W.:* Parathyroid function in human vitamin D deficiency and vitamin D deficiency in primary hyperparathyroidism. Am. J. Med. *56:* 833–839 (1974).
30 *Malluche, H.H.; Ritz, E.; Kutschera, J.; Krause, G.; Werner, E.; Gati, A.; Seifert, U., and Lange, H.P.:* Calcium metabolism and impaired mineralisation in various stages of renal insufficiency; in *Norman, Schaefer, Grigoleit, von Herrath and Ritz* Vitamin D and problems related to uremic bone diseases, pp. 513–522 (de Gruyter, Berlin 1975).
31 *Massry, S.G.; Arieff, A.I.; Coburn, J.W.; Palmieri, G., and Kleeman, C.R.:* Divalent ion metabolism in patients with acute renal failure: studies on mechanism of hypocalcaemia. Kidney int. *5:* 437–445 (1974).
32 *Massry, S.G.; Coburn, J.W.; Lee, D.B.N.; Jowsey, J., and Kleeman, C.R.:* Skeletal resistance to parathyroid hormone in renal failure. Ann. intern. Med. *78:* 357–364 (1973).
33 *Massry, S.G.; Dua, S.; Garty, J., and Friedler, R.M.:* Role of uremia in the skeletal resistance to the calcaemic action of parathyroid hormone; in Third Workshop on vitamin D (in press, 1977).
34 *Massry, S.G.; Stein, R.; Garty, J.; Arieff, A.I.; Coburn, J.W., and Norman, A.W.:* The role of 1,25-dihydroxycholecalciferol in the skeletal resistance to the calcaemic action of parathyroid extract; in *Norman, Schaefer, Grigoleit, von Herrath and Ritz* Vitamin D and problems related to uremic bone disease, pp. 619–622 (de Gruyter, Berlin 1975).
35 *Mawer, E.B.; Backhouse, J.; Davies, M., and Hill, L.F.:* Metabolic fate of administered 1,25-dihydroxycholecalciferol in controls and in patients with hypoparathyroidism. Lancet *i:* 1203–1206 (1975).
36 *Mawer, E.B.; Backhouse, J.; Holman, C.A.; Lumb, G.A., and Stanbury, S.W.:* The distribution and storage of vitamin D and its metabolites in human tissues. Clin. Sci. *43:* 413–431 (1972).
37 *Mawer, E.B.; Backhouse, J.; Taylor, C.; Lumb, G.A., and Stanbury, S.W.:* Failure of formation of 1,25-dihydroxycholecalciferol in chronic renal insufficiency. Lancet *i:* 626–628 (1973).
38 *Mawer, E.B.; Lumb, G.A.; Schaefer, K., and Stanbury, S.W.:* The metabolism of isotopically labelled vitamin D_3 in man: the influence of the state of vitamin D nutrition. Clin. Sci. *40:* 39–53 (1971).
39 *Mayer, G.P.:* Effect of calcium and magnesium on parathyroid hormone secretion in calves; in *Talmage, Owen and Parsons* Calcium-regulating hormones. Proc. Fifth Parathyroid Conf., Oxford, pp. 122–124 (Excerpta Medica, Amsterdam 1975).

40 *Parsons, L.G.:* The bone changes occurring in renal and coeliac infantilism and their relation to rickets. Archs Dis. Childh. *11:* 1–25 (1927).
41 *Popovtzer, M.M.; Matthay, R.; Alfrey, A.C.; Block, M.; Beck, P.; Miles, J., and Reeve, E.B.:* in *Frame, Parfitt and Duncan* Clinical aspects of metabolic bone disease, pp. 382–387 (Excerpta Medica, Amsterdam 1973).
42 *Sherrard, D.J.; Brickman, A.S.; Baylink, D.J.; Maloney, N.; Norman, A.W., and Coburn, J.W.:* Effects of $1,25(OH)_2D_3$ on bone disease of chronic renal failure; in Third Workshop on vitamin D (in press, 1977).
43 *Slatopolsky, E.; Rutherford, W.E.; Hoffsten, P.E.; Elkan, I.O.; Butcher, H.R., and Bricker, N.S.:* Non-suppressible secondary hyperparathyroidism in chronic progressive renal disease. Kidney int. *1:* 38–46 (1972).
44 *Stanbury, S.W.:* Azotaemic renal osteodystrophy. Br. med. Bull. *13:* 57–60 (1957).
45 *Stanbury, S.W.:* Bony complications of renal disease; in *Black* Renal disease; 2nd ed., pp. 665–713 (Blackwell, Oxford 1967).
46 *Stanbury, S.W.:* The intestinal absorption of calcium in normal adults, primary hyperparathyroidism and renal failure; in *Berlyne* Nutrition and renal disease, pp. 118–130 (Livingstone, Edinburgh 1968).
47 *Stanbury, S.W.:* Bone disease in uremia. Am. J. Med. *44:* 714–724 (1968).
48 *Stanbury, S.W.:* Azotaemic renal osteodystrophy. Clins Endocr. Metab. *1:* 267–304 (1972).
49 *Stanbury, S.W.:* The role of vitamin D in renal bone disease. Clin. Endocr., suppl. (in press, 1977).
50 *Stanbury, S.W. and Lumb, G.A.:* Metabolic studies of renal osteodystrophy. I. Calcium, phosphorus and nitrogen metabolism in rickets, osteomalacia and hyperparathyroidism complicating chronic uraemia and in the osteomalacia of the Fanconi syndrome. Medicine, Baltimore *41:* 1–31 (1962).
51 *Stanbury, S.W. and Lumb, G.A.:* Parathyroid function in chronic renal failure. Q. Jl Med. *35:* 1–23 (1966).
52 *Stanbury, S.W. and Lumb, G.A.:* The osteomalacia of renal insufficiency; in *Hioco* L'ostéomalacie, pp. 367–378 (Masson, Paris 1967).
53 *Stanbury, S.W. and Lumb, G.A.:* Parathyroid function in chronic vitamin D deficiency in man: a model for comparison with chronic renal failure. Calcif. Tiss. Res., suppl., pp. 185–201 (1976).
54 *Stanbury, S.W.; Lumb, G.A., and Mawer, E.B.:* Osteodystrophy developing spontaneously in the course of chronic renal failure. Archs intern. Med. *124:* 274–281 (1969).
55 *Stanbury, S.W.; Lumb, G.A., and Nicholson, W.F.:* Elective subtotal parathyroidectomy for renal hyperparathyroidism. Lancet *i:* 793–799 (1960).
56 *Stanbury, S.W.; Lumb, G.A., and Torkington, P.:* Interrelations of the parathyroid glands, the bone and the kidneys in human vitamin D deficiency; in *Hioco* Calcium et maladies endocriniennes, pp. 31–46 (Sandoz, Paris 1976).
57 *Vosik, M.; Anderson, C.F.; Steffee, W.P.; Johnson, W.J.; Arnaud, C.D., and Goldsmith, R.S.:* Successful medical management of osteitis fibrosa due to 'tertiary' hyperparathyroidism; Mayo Clin. Proc. *47:* 110–113 (1972).

Dr. *S.W. Stanbury*, University of Manchester, Department of Medicine, The Royal Infirmary, *Manchester M13 9WL* (England)

Subject Index